Social Lives of Plants

How plants connect
—as friends or foes—
with other plants, microorganisms, animals, and humans

SOCIAL LIVES OF PLANTS

Youngsook Lee • Baeyoung Choi

Youngsook Lee • Bae Young Choi

Social Lives of Plants

How plants connect—as friends or foes—with other plants, microorganisms, animals, and humans

Springer

Youngsook Lee
Department of Life Science
Pohang University of Science and Technology
Pohang, Gyeongsangbuk-do, Korea (Republic of)

Bae Young Choi
School of Liberal Arts and Sciences
Korea National University of Transportation
Chungju-si, Korea (Republic of)

ISBN 978-981-96-9836-3 ISBN 978-981-96-9837-0 (eBook)
https://doi.org/10.1007/978-981-96-9837-0

English translation of the 1st original Korean edition published by The EAST-ASIA Publishing Co., Seoul, 2024

This Springer imprint is published by the registered company Springer Nature Singapore Pte Ltd.
The registered company address is: 152 Beach Road, #21-01/04 Gateway East, Singapore 189721, Singapore

Foreword

Social interactions are essential for human survival. We play and work with other people, seek recognition and love, and obtain what we need to survive and thrive through social interactions. Similarly, social interactions are important for plants. But, how do plants, permanently rooted in one spot, engage in social activities? Why do plants need a social life? And why should we care about the social life of plants? This book aims to address these questions.

Plants constantly engage in social interactions with a wide range of other organisms, including not just other plants but also bacteria, viruses, fungi, insects, animals, and humans. The relationship between plants and other life forms may seem one-sided, since only plants can convert solar energy into sugars and wood, and other organisms depend on plants for food and shelter either directly or indirectly. Yet despite being immobile and surrounded by numerous organisms that would like to take advantage of them, plants successfully defend themselves and propagate. The dominance of green across Earth's surface is evidence of their success.

Plants do not merely endure or defend themselves—they also actively use other organisms for their growth and reproduction. The strategies plants employ for such interactions are astonishingly diverse and innovative. Plants establish symbiotic relationships with bacteria and fungi to acquire nutrients, and recruit insects for reproduction. Plants produce various toxic substances to ward off pathogens and herbivores, some of which have in turn evolved methods to detoxify these poisons. This aspect of a plant's social life resembles the arms race of modern nations, in which countries constantly develop new weapons in case other countries attack them. Through these interactions, plants and their social partners bring about changes in ecosystems and enrich them.

The social life of plants is also important for humans. We obtain food and shelter from plants. Moreover, by selectively breeding plants, we have developed crops that provide enormous amounts of calories and diverse materials. Our interactions with plants through the ages have allowed explosive human population growth and subsequent cultural development. However, as more people desired more and better things, Earth's natural systems were increasingly strained to meet human needs. This growing pressure has led to rapid environmental changes, such as acid rain, heavy metal pollution, and climate change. In other words, highly productive plants enabled humans to develop sophisticated civilizations, but human activities disrupt the environment, threatening the survival of many ecosystems.

In our rapidly changing environment, it is dangerous to assume that plants will continue to provide us with cheap foods and materials. Recent global weather changes often caused decreases in plant productivity, showing that such assumptions are ill-founded. As healthy plants are crucial for sustaining human life on Earth, it is critical that we study all aspects of plant life in detail and determine how to protect their environment and support their growth.

This book covers the social interactions of plants with other plants, microbes, and animals in Parts I–III. There are many scientific papers on these topics, which were referenced here, and further reading is suggested. Part IV discusses the interactions between plants and humans. Although papers have been published on this topic, we did not think they were sufficient to predict the future reliably. However, we tackled this topic to the best of our knowledge, believing that we should think about our future as a duty to our society.

In addition to explaining scientific facts, we pose many questions in this book. We explore which questions led scientists to discover certain facts and discuss what future research questions are necessary. In doing so, we hope to encourage active questioning, as people tend to accept what they learn passively but hesitate to ask questions. Asking questions clarifies what is already known and what remains unknown. Some might hesitate to ask questions, fearing that they may look foolish. However, seemingly trivial questions might lead to important questions and, subsequently, to important discoveries.

We also aim to explain the experimental methods that led to various conclusions. While we do not want to overwhelm non-expert readers by explaining the details, we would like to provide some basic explanations that might help readers understand how scientists approach problems. Besides ecological and physiological methods, we also discuss advancements in biochemistry, genetics, and phylogenetics, such as DNA sequencing and metabolomics, since recent developments in these technological areas have provided great insight into the relationships between plants and other life forms. Previously, one could only observe and describe interactions between organisms, and speculate about the meaning of these interactions, but now it is possible to identify exactly which chemicals and genes are involved in an interaction, and to propose plausible evolutionary processes caused by such interactions. We strived to explain the biochemical basis and evolutionary steps of the interactions. For readers interested in the technical details, a section at the end of Part IV elaborates on recent advancements in biochemistry, genetics, and phylogenetics. We also included footnotes for readers needing explanations of scientific terminology.

In this book, we describe how plants compete, cooperate, or exploit, but we do not intend to anthropomorphize plants and suggest that they act with intent. Rather, we followed conventional ways of expression familiar to most people. To be precise, we should say, "it is in a plant's nature to compete or to help one another," or "the results of long evolution to optimize a plant's survival in its individual niche made them compete, cooperate, or exploit." However, such expressions sound awkward and unnatural, so we opted for simpler and more natural descriptions.

Lastly, since we authors have studied plants in the laboratory only and not in the field, we feel that the ecological knowledge presented in this book might be incomplete. We kindly ask for your understanding and feedback. Although imperfect, we hope this book helps readers understand how plants live and interact with other life forms.

Pohang, Republic of Korea Youngsook Lee
Chungju, Republic of Korea Bae Young Choi

Contents

Part I

The Social Life of Plants with Their Neighboring Plants

How do stationary plants engage in social interactions with each other? Plants need light, water, and inorganic nutrients to survive. In most environments, these elements are not plentiful. Thus, competition seems to be a natural consequence of these conditions. But surprisingly, the opposite of competition has recently been observed; plants cooperate by exchanging nutrients with each other. At the same time, plants also closely cooperate with each other to fend off common enemies such as bacteria, fungi, and insects. Just as siblings compete and cooperate most intensely among humans, plants also seem to compete and cooperate more intensely with other plants than with any other organisms. In this part, we will first explore the cooperation among plants, and then examine the competition among them.

Cooperation Among Plants 1

Plants Are Interconnected and Help Each Other

Plants appear to stand separately. They look like they cannot embrace their friends or push away intruders. But are they really living separately? No. Recent research revealed that, in the unseen world beneath the ground, plants are intricately connected, forming large communities and helping each other (Fig. 1.1).

After learning that plants are interconnected and that this connection benefits them, trees suddenly appear quite different to me than before. The trees lining the street, visible through office windows, once seemed to stand independently, but now I can imagine them holding hands beneath the ground. Most of the trees in our

Fig. 1.1 Plants connected underground through mycorrhizae. Plants share nutrients and signals through fungal connections between their roots. (Image used with permission from *New Phytologist* [133])

Y. Lee, B. Y. Choi, *Social Lives of Plants*,
https://doi.org/10.1007/978-981-96-9837-0_1

Fig. 1.2 Fungal hyphae covering plant roots. Illustration of mycorrhizal symbiosis, with fungal hyphae spreading from plant roots into the soil, forming a vast underground network that aids nutrient uptake. (Image provided by Michael Rothman, who retains the copyright)

neighborhood are likely holding hands underground. The trees in my house are probably connected to the trees in the neighboring houses. My neighbor, a grizzled old man, diligently takes care of the trees in his garden every day. Perhaps, thanks to him, the trees in my house are also growing well. Although I have been a bit negligent, it is possible that my trees have thrived due to the diligence of my neighbor.

How are plants connected to other plants? What is the mechanism? Plants are physically connected to other plants through symbiotic relationships with mycorrhizal fungi. About 80–90% of terrestrial plants form these associations at their roots. This structure, where fungi colonize plant roots, is called "mycorrhiza," and the fungi that live symbiotically with plant roots are called "mycorrhizal fungi." Almost all plants in nature need the help of mycorrhizal fungi to grow well because environments with ample water and inorganic nutrients, which are essential for plant survival, are rare.

But how do fungi benefit plants? Mycorrhizal fungi grow by producing many long and thin threads called mycelia, which together create a very large surface area (Fig. 1.2). Through this extensive surface, these mycelia absorb water and inorganic nutrients, such as nitrate, potassium, and phosphate, dissolved in water from the soil. The fungi then provide the absorbed water and inorganic nutrients to the plants. In return, the plants supply carbohydrates and fats for the fungi to feed on and grow.

When transplanting trees, it is customary to dig up and move the soil around the trees' roots along with the roots themselves (as shown in Fig. 1.3). This is because the soil contains mycorrhizal fungi that live in symbiosis with the tree, and when the fungi are moved together with the tree, it helps the tree adapt more easily to a new environment.

The mycorrhizal symbiosis for nutrient exchange between plants and fungi has been known since the twentieth century. However, toward the end of the twentieth century, scientists discovered that the role of mycorrhizal fungi does not stop there; as seen in Fig. 1.1, *the mycelia of mycorrhizal fungi wrap around the roots of many plants and connect them, forming a plant community*. Scientists have reported that this plant community shares signals of plant wounds or pathogen attacks through the mycorrhizal fungi's mycelia, enabling multiple plants to defend against

Fig. 1.3 A 20-year-old Japanese red pine (*Pinus densiflora*) is transplanted to a new location along with the surrounding soil. When transplanting a large tree, it is important to move the surrounding soil as well to ensure the transfer of symbiotic mycorrhizal fungi, which help the tree adapt to and thrive in its new environment

enemies together. They have also found many lines of evidence that a photosynthesizing tree shares sugars with a tree that cannot photosynthesize, via the mycorrhizal connection.

Let us discuss which experiment scientists conducted to discover that plants help each other. The first group to study this topic in a natural environment were Dr. Simard's research team in Canada [120]. They found that seven mycorrhizal fungi species covered the root tips of both the paper birch (*Betula papyrifera*) and Douglas fir (*Pseudotsuga menziesii*) and suspected that these mycorrhizal fungi connected the two trees, allowing them to share nutrients. To test this possibility, they performed isotope[1] feeding experiments. They fed (through leaves) the paper birch with carbon dioxide labeled with ^{13}C and the Douglas fir with carbon dioxide labeled with ^{14}C, allowing them to photosynthesize. As a result, the paper birch produced carbohydrates labeled with ^{13}C, and the Douglas fir produced carbohydrates labeled with ^{14}C. But if these trees were connected and sharing nutrients, would not carbohydrates produced by one tree be detected in the other? Exactly. This was the most important question they wanted to ask. Indeed, a small amount of carbohydrates labeled with ^{14}C was detected in the paper birch, and a small amount of carbohydrates labeled with ^{13}C was detected in the Douglas fir (Fig. 1.4). This means that the photosynthetic products moved bidirectionally. They found out that between 3% and 10% of the total photosynthesized amount was transferred.

Taking a step further, the scientists investigated whether one tree could provide carbohydrates to another tree that could not photosynthesize. When they shaded the

[1]The mass of an atom (atomic mass) is determined by the sum of the protons and neutrons in its nucleus. In the case of isotopes, the number of protons remains the same, but the number of neutrons differs, resulting in a variation in atomic mass. For example, more than 98% of naturally occurring carbon consists of atoms with 6 protons and 6 neutrons, giving it an atomic mass of 12 (denoted as ^{12}C). However, the carbon isotope ^{13}C has 6 protons and 7 neutrons, resulting in an atomic mass of 13. Since isotopes exist in very low natural abundances, they can be used as tracers to track the movement and transformation of substances.

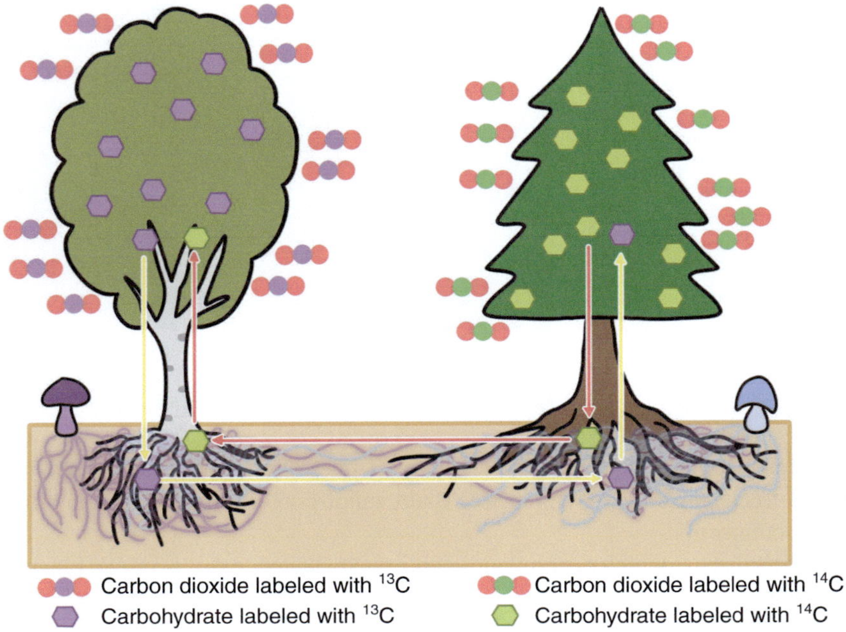

Fig. 1.4 An illustration explaining an experiment that revealed the exchange of carbohydrates between a paper birch and a Douglas fir. The paper birch (on the left) was fed carbon dioxide labeled with ^{13}C (purple circles), and the Douglas fir (on the right) was fed carbon dioxide labeled with ^{14}C (green circles), to be used in photosynthesis (red circles are oxygens). As a result, the paper birch produced carbohydrates labeled with ^{13}C (purple hexagons), and the Douglas fir produced carbohydrates labeled with ^{14}C (green hexagons). However, a small amount of carbohydrates labeled with ^{14}C was detected in the paper birch, and carbohydrates labeled with ^{13}C were detected in the Douglas fir. This shows that the two trees exchanged photosynthetic products. The researchers speculated that this exchange occurred through the fungal hyphae connecting the two trees

Douglas fir to block sunlight, they observed a significantly greater amount of carbohydrates moving from the paper birch to the fir. The amount of carbohydrates was substantial enough to help the plant survive under harsh environmental conditions. These types of experiments revealed that even trees of different species, if connected by mycorrhizal fungi, shared nutrients, enabling them to better withstand challenging conditions such as drought or nutrient deficiency.

Then, does the connection between trees change when environmental conditions change? Multiple studies, including those by the Simard research team, revealed that the mycorrhizal networks that connect plants change depending on environmental conditions and the developmental stages of the plants [121]. For instance, trees strengthen their mycorrhizal connections with neighboring trees when environmental conditions deteriorate. Under stressful conditions with very low soil moisture, the mycorrhizal networks between drought-resistant and drought-sensitive plants become tighter. Through these networks, nutrients are transferred from the resistant plants to the sensitive plants, helping the latter survive the drought.

Another example is young plants growing beside mature trees. The young plants start their mycorrhizal networks by tightly connecting to the already established mycorrhizae of the mature trees. Young plants that are well connected to the mycorrhizal networks of mature trees can obtain many nutrients necessary for their initial growth. There have also been experimental results showing that dying trees transfer nutrients to young trees. In a 2015 experiment, scientists found that lodgepole pine trees, severely stressed by leaf removal, transferred carbohydrates and stress signals to young ponderosa pines through mycorrhizal hyphae [122]. However, when a very fine mesh was placed to prevent the hyphae from connecting the two trees, no transfer of carbohydrates and stress signals was observed. This indicates that the mycorrhizal hyphae were the passageway for the carbohydrates and stress signals between the two trees. At the experimental site, many lodgepole pines were damaged by drought and pests, worsened by global warming, while ponderosa pines, being a species that can survive better under such climate conditions, were gradually replacing the lodgepole pine forest. One might imagine from this result that "dying lodgepole pines pass on what they have to young ponderosa pines to help them grow quickly, as if plants also leave a legacy." However, the authors, being cautious scientists, explained in the paper that "the rapidly growing ponderosa pines absorb a large amount of nutrients, lowering the concentration of inorganic nutrients in the soil. As a result, due to the difference in nutrient concentration in the soil, diffusion occurred through the mycorrhizal network, transferring nutrients from the lodgepole pines to the ponderosa pines. There is a possibility that the mycorrhizal fungi play a role in the transition from the lodgepole pine forest to a ponderosa pine forest."

Through various experimental results, people have come to understand that plants share materials through their interconnected roots, living as if they are one large organism. While this finding is indeed amazing and remarkable, I would like to suggest that we take one more step to ask further scientific questions: consider the opposite, raise questions, and propose what needs to be studied further. Because doing this will help us better understand this fascinating phenomenon. Are all nearby plants always connected? Are there groups of plants that connect better with each other? Are there any plant groups that prefer a solitary way of living? Are carnivorous plants, which obtain nutrients by catching insects, less connected to other plants? Is the mycorrhizal network used solely for mutual helping? Could it be used to suppress the growth of neighboring plants? Although there are not many papers on these topics yet, if more people take an interest in the social life of plants, such research will increase, leading to a better understanding of these phenomena.

All Plants Stand Guard Against Pests and Diseases

Plants cooperate with each other in defense against pests and diseases. Under normal conditions, plants focus on growth, but when attacked by pathogens or insects, they switch to defense mode. A plant that fails to rapidly change to defense mode when pathogens attack is likely to succumb to the attack and die. So, how do plants quickly detect the presence of pests or diseases? The answer is in detecting signals from neighboring plants; rather than waiting until it is actually attacked, a plant can detect when a neighboring plant is being attacked, prepare its defenses in advance, and respond more effectively to the threat.

In 1997, the research group of Professor Raskin at Rutgers University in the United States reported experimental evidence that plants can detect when neighboring plants are under attack and prepare their defenses in advance [119]. It was speculated at that time that salicylic acid (SA) played an important role in the plants' defense response to pathogens. To find out how SA might work, they investigated the distribution and movement of SA inside tobacco plants infected with the tobacco mosaic virus. Their purpose might have been to test whether SA was a signal moving inside the plant to alert the whole plant body that the pathogen had arrived, but instead, they found that methyl salicylate (MeSA), which can be synthesized from SA, was released into the air from the virus-infected tobacco plants. MeSA is a volatile compound with a minty scent, commonly used in wintergreen oil and as a flavoring agent.

To investigate the role of MeSA released from virus-infected leaves in plant virus resistance, they placed tobacco plants in a sealed container to prevent gas from escaping and added MeSA to the container. Six days after the treatment, they observed that the concentration of SA in the tobacco leaves had increased many hundredfold, and the expression of genes[2] involved in disease resistance was elevated. To determine if resistance to the virus had changed, they infected the MeSA-treated tobacco leaves with the virus, and measured the area of the leaves infected by the virus. Indeed, the infected area was smaller in MeSA-treated plants than in the non-treated plants, and the difference was more pronounced with higher concentrations of added MeSA. These results suggested that MeSA could act as a signaling molecule to activate plant defense mechanisms against viruses.

In the next step, they conducted an experiment to see if the concentration of MeSA released by the virus-attacked plants was high enough to affect healthy neighboring plants. As shown in Fig. 1.5, a tobacco plant infected with the virus (A) was placed in a container, and the air inside was connected through a glass tube to allow it to be transferred to a healthy "neighbor" tobacco plant (B). As a control for this experiment (experiment I), they repeated the experiment (experiment II), but inserted a filter into the connecting tube to prevent MeSA from reaching the healthy tobacco plant B. After 6 days, it was found that the gene *PR-1*, important for the defense response, was expressed at much higher levels in the leaves of plant B in experiment I compared to that of plant B in the control experiment II. In addition, when the virus was inoculated into plant B of both experiments, the infected area in experiment I was much smaller compared to the infected area of the control experiment II. Based on these results, they hypothesized that the concentration of MeSA released into the air by the virus-infected plant A was sufficient to activate the

[2] A gene is a segment of DNA that carries the instructions for making proteins. Genes determine an organism's traits and are passed from parents to offspring.

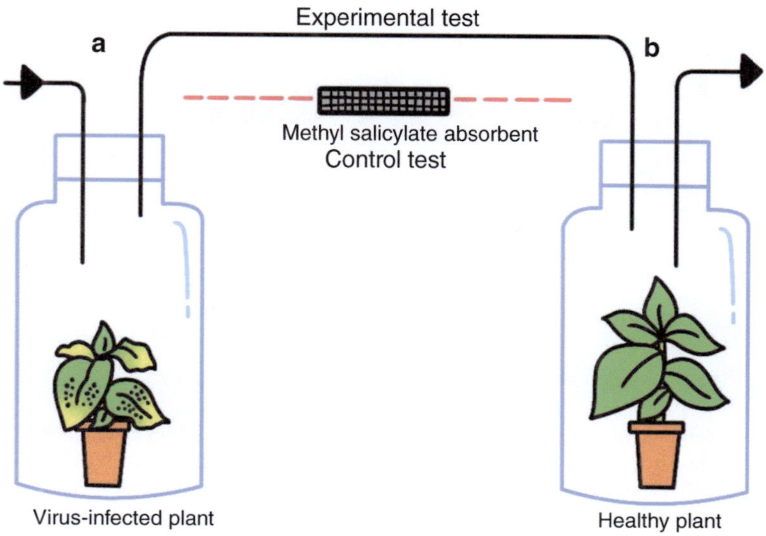

Fig. 1.5 A schematic design of the experiment that investigated whether a healthy plant could detect volatile compounds emitted by a virus-infected plant and initiate a defense response. Two glass jars, each containing a plant, were connected with a tube, allowing gas to move between them. The jar on the left (**a**) contained a virus-infected plant, while the one on the right (**b**) contained a healthy plant. Pressure was applied to ensure the gas moved from the jar on the left to the one on the right, allowing the volatile compounds released by the infected plant to be absorbed by the healthy plant. When analyzed after six days, the healthy plant expressed defense-related genes. Moreover, when artificially infected with the virus, its resistance to the virus was improved. However, as a control, when a filter that absorbs methyl salicylate (MeSA) was placed between the jars to prevent MeSA from reaching the healthy plant, the healthy plant did not show enhanced virus resistance

defense genes of neighboring plants B, thereby improving their resistance to the virus.

Since then, many scientists have conducted studies on how volatile substances emitted by plants enhance plant resistance to pests and diseases. These experiments revealed that when a plant becomes diseased or is attacked by insects, it sends out a warning signal to the surrounding plants. The warning signal consists of various aromatic compounds emitted by the plant when it is wounded. When these aromatic compounds disperse into the air, nearby plants that have not yet been attacked can "smell" them and switch from a growth mode to a defense mode, reducing damage when attacked (Fig. 1.6). Thus, there seem to be no designated sentinels in the plant community—any plant attacked plays the role of a sentinel. For this reason, people conducting experiments with plants are often cautious about wearing perfume, as the aromatic compounds in perfume could be mistaken by plants for distress signals from their neighbors, potentially triggering a defense response and thereby affecting the experiment.

The distress signals that plants emit, such as the aforementioned aromatic compounds, disperse into the air and are perceived not only by plants of the same

Fig. 1.6 Volatile alarm signals from pest-attacked plants trigger defenses in neighbors. When a plant is attacked by pests or pathogens, it releases volatile compounds that act as alarm signals. Nearby plants detect these signals and preemptively activate their own defense mechanisms, even before being attacked themselves

species but also by plants of different species. As gaseous signals, while fast, dissipate quickly and do not stay for long, signals other than gaseous compounds are most likely also being shared. Plant defense hormones, for example, may be directly transmitted through the fungal hyphae from the attacked plant to other plants, when mycorrhizal fungi connect their roots.

The idea that plants communicate with neighboring plants is both fascinating and intriguing. This phenomenon could, of course, be described not as the wounded plant actively sending signals but rather as the neighboring plants noticing or "eavesdropping" on what is happening to the wounded plant. However, one thing is certain: information spreads among plants.

The evolution of information-sharing mechanisms, which allow plants to detect when neighboring plants are under attack, must have conferred a significant advantage to plants. However, pathogens have also developed new methods to continue invading plants. Just like in an arms race, plants and pathogens continuously develop new ways to overcome the invasion and defense mechanisms of their opponents, resulting in the evolution of complex, multilayered strategies of attack and defense.

Plants That Act as Foster Parents

The *Carnegiea gigantea*, commonly called the saguaro cactus, is the largest cactus in the United States. It grows to over 10 meters tall and has a diameter of about 75 centimeters, providing food for much wildlife. This large cactus requires a "foster parent" in its early stages of growth, because after germination, the very young cactus can easily dry out and die without water. Therefore, young saguaros can only grow under large trees that can absorb water from deep in the ground. Large trees from the legume family, such as *Parkinsonia microphylla*, play the role of a "foster

Fig. 1.7 A Saguaro cactus that matured under the care of its nurse plant, the Palo Verde. Now similar in height, the Saguaro initially developed under the Palo Verde's canopy, which provided shade, moderated temperature, and enhanced soil nitrogen through symbiosis with nitrogen-fixing bacteria. (Photo by salvmell, via iNaturalist. Licensed under CC BY 4.0)

parent" for the saguaro (Fig. 1.7). Under the shade of the foster parent tree, the saguaro is protected from the heat of summer, drought, excessive sunlight, and the cold of winter, while also obtaining nitrogen and water.

Saguaros grow extremely slowly, taking 20–50 years to reach a height of 1 meter. Eventually, when the saguaro grows larger than its foster parent tree, it absorbs so much water and so many nutrients from the soil that the parent tree often dies from a lack of nutrients. This is why young saguaro cacti grow under large trees, but there are few large plants next to fully grown saguaros. Some scholars also argue that species like the Norway spruce and larch serve as foster parents for broad-leaved trees.

Scientists do not claim that "foster parent" trees intentionally nurture young plants or sacrifice themselves. Instead, they suggest that young saguaros or broad-leaved trees may have simply found themselves, by chance, in favorable environments under large trees and taken advantage of the conditions. As the young trees grow vigorously, the foster parent trees may struggle to obtain the necessary resources, eventually dying and being replaced by the younger trees. Scientists interpret this phenomenon as a form of ecological succession,[3] where growing trees

[3] Ecological succession refers to the process by which the composition of a biological community changes over time due to various ecological factors such as environmental changes, nutrient availability, competition, etc.

alter the environment, making it more suitable for new plant species than for the older trees. This natural process leads to changes in the species composition of the forest.

Can Plants Recognize Their Kin?

Some scientists argue that plants recognize their kin, cooperate with them, and share resources. If true, this would mean plants have social interactions similar to those in humans; people feel closer to their relatives and tend to help them more. Although not many experiments have been conducted on plant kin recognition, and the topic has not yet attracted much attention in the scientific community, we will explore the experiments that have led some scientists to make such claims since it is a core question of this book.

In 2007, Professor Dudley from Canada reported that *Cakile edentula*, a succulent plant found on beaches, behaved differently depending on the plants around it. When surrounded by plants of the same species, it shared space while establishing roots. However, when a different species was nearby, it competed aggressively [33]. This finding suggests that plants may be able to distinguish between genetically similar and different individuals. However, many scholars raised concerns, pointing out flaws in the experiment. In response, other researchers have since investigated whether such behavior actually occurs.

Dr. Torices from Spain discovered evidence of kin recognition in the flowering behavior of *Moricandia moricandioides*, a fragrant plant [130]. The experiment took place in a greenhouse, where seeds were planted in pots, and the size and number of flowers were later counted. To test for kin recognition, sibling seeds from the same plant were placed in one pot, while seeds from different plants in the population were placed in another. In each pot, the test seeds were planted in the center, surrounded by either sibling seeds or, in the control group, non-sibling seeds. When the plants flowered, it was found that those surrounded by siblings produced larger petals and more flowers than those in the control group.

Here is Dr. Torices' interpretation of the results: larger and more numerous flowers create a "magnet effect," attracting more insects to spread pollen. When *Moricandia* produces bigger, more abundant flowers, it is more likely to attract bees, improving its chances of successful pollination. However, if a plant is isolated and produces large, showy flowers, it might waste its energy on flower production, leaving fewer resources for seed production, which could hurt its reproductive success. *Moricandia* has evolved to recognize its kin. When growing near relatives, it produces more and larger flowers, boosting the reproductive success of the entire group.

Dr. López Pereira from Argentina studied sunflowers that produced only a single stem without side branches. When planted closely together in tight rows, the sunflower stems grew at an angle, as if allowing nearby plants to access light, and all of them grew reasonably well [78]. Using this phenomenon, sunflowers could be planted very densely, boosting sunflower oil production per unit area by up to 47%.

Further research revealed that the sunflowers adjusted their stem angle in response to light reflected from neighboring plants. The reflected light was later found to be detected by a pigment-protein in plant, called phytochrome.

Some scientists suggest that trees connected through dense networks of mycorrhizal fungi may send more nutrients and chemical signals to trees of the same species. Dr. Pickles from the United Kingdom discovered through experiments that *Pseudotsuga menziesii* (Douglas fir) trees provided more nutrients to young Douglas firs than to young trees of other species, and they also sent stronger signals when under insect attack [101]. This raises the question: could forests made up mostly of a single tree species, like Douglas fir, exist because these trees recognize and support their kin? In Korea, there are beautiful fir forests in places such as Gwangneung Arboretum, Woljeongsa Temple, and Naesosa Temple. It would be interesting to investigate whether fir trees also recognize and support their kin, thus forming fir forests.

Dr. Semchenko's team in Estonia conducted research on whether plants can recognize their kin by detecting substances secreted from the roots of neighboring plants [115]. The study used Tufted hairgrass (*Deschampsia cespitosa*), a tall grass species. To prevent the influence of soil nutrient depletion that might occur if Tufted hairgrass plants were grown directly next to each other, they were grown separately. The researchers extracted only the root secretions and applied them to the roots of other Tufted hairgrass plants. It turned out that when root secretions from an unrelated Tufted hairgrass were applied, the roots grew longer compared to when secretions from sibling plants were applied. Regarding this outcome, Dr. Semchenko stated, "When Tufted hairgrass detects the root secretions of an unrelated plant nearby, it develops longer roots to competitively acquire the limited resources in the soil. In contrast, when a sibling plant is nearby, it suppresses root development to avoid excessive competition." In short, there is no need for siblings to waste energy by excessively developing roots and competing. If this experimental result is accurate, it suggests that the types of chemicals secreted by plant roots differ even among individuals of the same species and that plants can recognize these subtle differences to distinguish their kin. However, Dr. Semchenko's team was unable to identify the specific chemical used to differentiate between sibling plants and unrelated plants of the same species. If the identity of this chemical could be uncovered, their claim would become more convincing.

The scientists who conducted these experiments claimed that plants recognize and cooperate with their kin. However, there are still many issues that need to be resolved before we can accept this idea as true. For example, can other researchers get the same results when conducting the same experiment? Are there alternative ways to explain this phenomenon? Are these general phenomena found across many plants, or are they rare exceptions? These questions need to be addressed first. Once these issues are resolved, the next step would be to understand *how* plants achieve this—how they recognize whether a neighboring plant is a sibling, another individual of the same species, or a member of a different species. There are many specific questions we can ask. Do plants detect physical or chemical signals from their neighbors for such recognition? Are these signals released from the roots or the shoots? How do they detect substances secreted by roots or volatile compounds emitted from shoots? Do plants possess specialized receptor proteins to recognize these signals? After recognizing their kin, how do plants assist them? If numerous experiments repeatedly produce solid answers to these questions, then we will be able to believe that plants truly recognize and cooperate with their kin. Scientists have the duty to uncover the truth. To do so, they must remain skeptical. Even if an explanation seems plausible, they must carefully examine whether the evidence is truly reliable.

Molecular Evidence That Plants Distinguish Between Kin and Nonkin

While the evidence for plants distinguishing between kin and nonkin through their roots and leaves is still insufficient, clear evidence has been found in the process where a stigma of a flower recognizes whether pollen comes from kin or a completely unrelated plant. This discovery was made because the topic of pollen recognition has been extensively studied for a long time by many scientists, due to its importance in improving agricultural yields. The accumulation of experimental results has led to solid evidence on this topic.

About 40% of flowering plants have systems in place to prevent self-pollination,[4] which is analogous to human customs or laws that prohibit close relatives from marrying and having offspring. These systems prevent a plant's own pollen, or pollen from genetically similar individuals, from fertilizing its egg cells, allowing only genetically different pollen to be accepted. This promotes the production of diverse offspring. If a plant were to self-pollinate, its offspring would be genetically identical or very similar, leading to reduced genetic diversity. Lower genetic diversity decreases the likelihood of survival when the environment becomes antagonistic to the plant. In contrast, if genetic diversity is high, there are likely to be individuals who can survive under new stressful conditions. Thus, many flowering plants have evolved mechanisms to prevent self-pollination. The process involves a protein in the stigma recognizing a protein in the pollen to determine whether the pollen comes from the plant itself, a relative, or an unrelated plant.

In contrast, in agriculture, it can be advantageous to disable these self-incompatibility systems. This is because if a crop has excellent traits,[5] self-pollination guarantees that these traits are passed on to the next generation without changes, allowing farmers to grow high-quality crops consistently. Because of this, research on how plants prevent self-pollination has been very active. As a result,

[4] Self-pollination refers to the process in which a plant allows its own pollen to germinate and grow on its own stigma, enabling the fusion of male gametes within the pollen and female gametes within the pistil of the same plant to produce offspring.

[5] A trait is a characteristic of a living organism, such as its physical appearance, biochemical makeup, physiological function, or behavior.

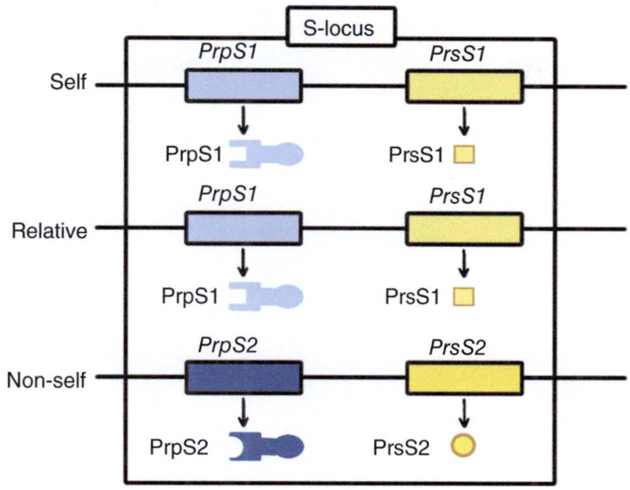

Fig. 1.8 A schematic diagram explaining the set of self-incompatibility genes located at the S-locus in plants of the poppy family. The S-locus codes for proteins expressed in the pistil and pollen that recognize each other. The *PrpS* gene is expressed in the pollen, and the *PrsS* gene is expressed in the pistil, producing the proteins PrpS and PrsS, respectively. Since the *PrpS* and *PrsS* gene set is located at the same position on the chromosome, they are always inherited together, enabling the plant to distinguish between self and nonself, including relatives

scientists have discovered many ways plants use to recognize whether pollen on their stigma is from themselves or from another plant at the molecular level.

To explain how pollen is recognized on the stigma, we first need to talk about the S-locus. The prevention of self-pollination is controlled by a group of genes found in a part of the genome called "the S-locus." These genes include a marker protein in the pollen and a matching receptor protein in the stigma that recognizes it. Since these two genes are in the same locus on the chromosome,[6] they are always passed down together and do not separate during reproduction (Fig. 1.8). The S-locus gene set is diverse—relatives have the same set, while unrelated plants have different sets.

How do the genes located at the S-locus prevent self-pollination? In plants of the poppy family, the genes at the S-locus express the PrsS protein in the stigma and the PrpS receptor protein in the pollen. These PrsS and PrpS proteins coming from the same S-locus, bind very specifically to each other, like a lock and key. When a plant's own pollen—or pollen from a genetically similar relative with the same S-locus genes—lands on the stigma, PrsS binds to PrpS (Fig. 1.9a). Once PrsS binds to PrpS, the PrpS protein opens, allowing a large flow of calcium ions into the pollen, much like a levee breaking. This influx of calcium ions triggers a signaling

[6]A chromosome is a thread-like structure made of DNA that carries genetic information. During cell division, DNA becomes tightly packed into chromosomes to ensure the accurate transmission of genetic information to new cells.

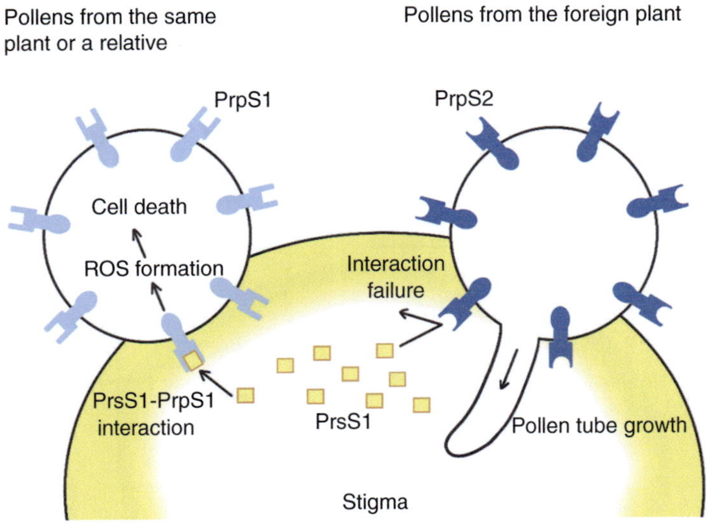

Fig. 1.9 The mechanism by which the pistil of a poppy flower prevents the survival of its own pollen or that of close relatives, while allowing foreign pollen to grow, fertilize the egg cell (eventually resulting in hybrid seeds). In a single plant and its relative, the same S-locus gene set is expressed in both the pollen and pistil. The PrsS1 protein in the stigma binds with the PrpS1 protein in the pollen, causing the pollen to die. In contrast, when foreign pollen lands on the stigma, the S-locus gene set is different, so PrsS1 does not bind with PrpS2. As a result, the pollen survives, grows a pollen tube, and fertilizes the egg cell

process[7] inside the pollen, leading to the production of a large amount of reactive oxygen species (ROS),[8] which ultimately causes the pollen to die. However, when pollen from a plant with different S-locus genes lands on the stigma, the PrsS protein in the stigma cannot bind to the PrpS receptor protein in the pollen (Fig. 1.9b). As a result, the pollen avoids cell death, grows a pollen tube, and successfully fertilizes the plant, leading to the production of offspring.

Unlike poppies, *Arabidopsis thaliana*, a weed plant often used in laboratories, does not prevent self-pollination, meaning that its pollen can fertilize egg cells in its own flower. However, when the *PrsS* gene was artificially expressed in the stigma

[7] A signaling process refers to the process by which a cell recognizes external stimuli (chemical or physical changes), transmits these changes internally, and triggers a series of cellular responses. A representative example is the restoration of water balance in cells. When a decrease in cellular water content is detected, the cell activates various signaling pathways to regain water balance. During the signaling process, secondary messengers are often generated, and changes in gene expression frequently accompany the process.

[8] Reactive oxygen species (ROS) are highly reactive molecules derived from oxygen, such as hydrogen peroxide. In biological systems, ROS have a dual role. At controlled levels, they help maintain cellular balance and act as signaling molecules in various processes. However, when ROS levels become excessive, they can damage vital cellular components, including DNA, proteins, and lipids. Accumulation of such damage can ultimately lead to cell death.

and the *PrpS* gene in the pollen of *Arabidopsis thaliana* using genetic engineering, the plant no longer accepted its own pollen [75]. Simply by expressing these two genes, self-pollination was successfully prevented.

Scientists who breed crops to improve their traits often try to stop self-pollination and create many hybrids, from which they can choose the ones with the best traits. If these two genes could be used to stop self-pollination in crops, it would become possible to produce diverse offspring and make the process of improving crops much easier.

As described above, plants in the poppy family prevent self-pollination by recognizing pollen with the same S-locus proteins and killing it. However, plants in the Solanaceae family (such as eggplant, tomato, and potato) prevent self-pollination differently—by recognizing only non-self-pollen and allowing it to survive. Scientists studying the evolution of self-pollination prevention in flowering plants have discovered that plants have developed at least 35 independent methods to prevent self-pollination.

With this kind of evidence, even the most skeptical scientists can no longer doubt that plants, at least in their reproductive organs, can distinguish between kin and nonkin. But what about other organs? While self-recognition in the reproductive organs is crucial for species survival—making it logical for plants to develop this ability—we can also imagine similar recognition occurring in other parts of the plant. If it can happen in one organ, it suggests that it might occur in other organs as well, especially if it provides a survival advantage to the plant.

Competition Among Plants

2

Competition for Light

The life of plants depends on light. They absorb light and use its energy for photosynthesis,[1] producing carbohydrates. These carbohydrates, along with the energy they contain, are used to create various chemical substances that allow the plants to grow. Thus, plants cannot survive in places without light, and in areas with insufficient light, they compete for it. Plants that grow in the shade are taller than plants of the same species that grow in sunlight (Fig. 2.1). This is because, to survive, they must rapidly reach a sunny spot and begin photosynthesis before their stored nutrients are depleted.

In nature, the main reason plants lack access to light is that other plants above them absorb it. But how do plants detect the presence of another plant above them? Could plants have something like eyes? Just as our eyes detect light through pigments that absorb it, plant leaves also contain various pigments that allow them to sense light. So, which color of light do plants sense to tell whether they are in the shade or directly exposed to sunlight? It is the red color. Since plant leaves absorb a lot of red light for photosynthesis, if the leaves above them absorb the red light, the leaves below will receive relatively less. By detecting how much red light is available, plants can determine if they are in the shade. Phytochrome, a bluish-colored protein that absorbs red light, plays a key role in this process.

When phytochrome is first produced in the plant, it is in the inactive Pr form. However, when this Pr form of phytochrome absorbs red light, it converts into the Pfr form, which triggers various changes in the plant (Fig. 2.2). The Pfr form is converted back to Pr form by absorbing far-red light. There is a higher amount of Pfr phytochrome in areas of plants with sufficient red light, while the Pr form is more

[1] Photosynthesis is the process of converting atmospheric carbon dioxide and water into sugar using light energy. During this process, light energy is transformed into chemical energy. The chemical energy is consumed to supply the energy necessary for biological activities.

Y. Lee, B. Y. Choi, *Social Lives of Plants*,
https://doi.org/10.1007/978-981-96-9837-0_2

Fig. 2.1 A photo comparing barley grown in bright light with barley grown in the shade. After germinating both pots of barley side by side in the light, the pot on the left was kept in the light, while the pot on the right was moved to the shade. The photo was taken two days later. The young barley grown in the shade in the pot on the right had narrower leaves, a lighter color, and was slightly taller than barley grown in the light. Of course, aside from the light, there may have been slight differences in humidity, temperature, and air movement between the two environments, which could also have affected their appearance. Scientists conducting experiments try to keep the conditions of the control group and experimental group as similar as possible, repeat the experiment multiple times, quantify the results, and investigate the statistical significance of any differences, to avoid drawing incorrect conclusions

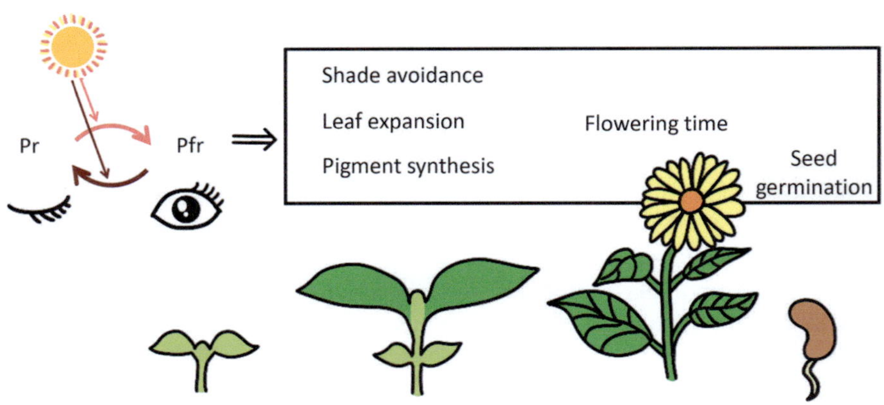

Fig. 2.2 A diagram explaining the mechanism by which plants sense and respond to light. The pigment called phytochrome in plants absorbs red or far-red light to help the plant recognize its light environment. When phytochrome absorbs red light (around 660 nanometers), it becomes the active Pfr-phytochrome form. When it absorbs far-red light (light with a wavelength of 700–800 nanometers, located at the far end of the visible red in the light spectrum), it becomes the inactive Pr-phytochrome form. The activated Pfr-phytochrome regulates various light-related developmental processes, such as shade avoidance, germination, and flowering

abundant in shaded areas. Pfr phytochrome enters the cell nucleus[2] and regulates the expression of various genes, breaking down proteins that inhibit light responses and thereby triggering a range of reactions to light. About 10% of the plant's genes change their expression depending on the state of phytochrome, highlighting the broad range of functions this pigment-protein controls. In addition to helping plants avoid shade, phytochrome regulates many other phenomena, including causing leaves to expand, synthesizing chlorophyll for photosynthesis in the presence of light, detecting seasonal changes to control the timing of flowering, and promoting seed germination.

Let us take a closer look at the shade-avoidance response regulated by phytochrome. When a plant is shaded by a neighboring plant, the neighbor absorbs the red light that best suits photosynthesis, reducing the amount of red light the shaded plant can receive. Since far-red light, which cannot be used for photosynthesis, is not absorbed by the neighboring plant, its amount remains the same with or without a neighboring plant. As a result, the ratio of red to far-red light decreases when a plant is shaded by a neighboring plant. When phytochrome detects this light environment, the plant's stem begins to grow longer, allowing it to escape the situation. This situation can be simulated by artificially irradiating the plant with far-red light, causing the stem to begin elongating within just 10 min. If the plant is still unable to escape the shade, it will flower earlier. However, such a plant produces fewer seeds, and many of the seeds it produces are unable to germinate since a plant that grows in the shade cannot perform sufficient photosynthesis and has limited resources to provide to the seeds; thus, the seeds are undernourished.

On the other hand, some plants are adapted to live in permanent shade. These shade plants do not elongate their stems in the shade. Instead, they produce thin, wide leaves and grow many side branches to capture as much light as possible. Adapting in this way is better for their survival than competing when the neighboring plant is a very tall tree, making competition very difficult.

The light conditions that plants are exposed to not only affect their growth but also have a significant impact on their resistance to pathogens and pests. In plants growing densely packed together, red light is absorbed by surrounding plants, leaving them with relatively more far-red light. In plants that receive a lot of far-red light, most of the phytochrome remains in its inactive state. Plants with inactive phytochrome are less able to activate their defense mechanisms against pathogens and pests, making them more susceptible to diseases and infestations [9]. A typical example of how plant density affects resistance to pests and diseases can be seen in crops. Farmers plant crops closely together to maximize yield from limited farmland. However, densely planted crops receive a large amount of far-red light, which lowers their resistance to pests and diseases, leading farmers to use pesticides. To address this issue, recent research has focused on using LEDs to artificially supplement red light, thereby enhancing crop resistance to pests and diseases [39].

[2]The nucleus is an organelle within the cell, surrounded by a lipid membrane, and it contains genetic material (DNA).

In 2006, researchers studying phytochrome at Jeju National University in South Korea developed a slow-growing grass variety by modifying the phytochrome gene. They eliminated the shade-avoidance response caused by phytochrome, which normally leads to rapid grass growth as it competes for light. When grass grows quickly, it needs to be mowed frequently, but since this modified grass no longer exhibits the shade-avoidance response, it grows slowly, making it easier to maintain. The inventors humorously suggested that by planting this grass, wives would nag their husbands less about mowing the lawn, potentially reducing marital arguments. For people in Europe and North America, where maintaining lawns is a common, labor-intensive household chore, this grass could be quite appealing by reducing the time spent mowing. However, due to concerns about the safety of genetically modified organisms (GMOs),[3] it has not received approval for commercialization.

An Invasive Species That Blocks Light and Displaces Native Species

The plant *Miconia calvescens* has large leaves, up to 1 meter long, that block light, preventing neighboring plants from growing (Fig. 2.3). Originally native to Mexico and Central and South America, *Miconia* has beautiful green or purple leaves with white veins, making it an attractive ornamental plant that was initially imported and planted in gardens. However, birds that ate the sweet *Miconia* fruit spread the seeds through their droppings, causing *Miconia* to grow in the wild.

One *Miconia* plant can produce as many as ten million seeds, and when these seeds sprout and grow large leaves, other nearby plants cannot receive enough light and are unable to grow. As a result, about 25% of the forests in Tahiti have turned into *Miconia* forests, and the plant has spread so widely in Sri Lanka and Hawaii that it has become a significant problem. While *Miconia* used its light-blocking competitive strategy to dominate new environments, native species had never coexisted with such a plant before and had no way to respond to it, failing in the competition and surrendering their habitats.

However, even the powerful *Miconia* has a weakness. By focusing on growing large leaves to outcompete other plants, *Miconia* fails to establish deep roots in the soil. Its shallow root system cannot firmly anchor the surrounding soil, leading to landslides during heavy rains in *Miconia* forests. Due to its rapid spread and role in causing landslides, the people of Tahiti refer to *Miconia* as the "green cancer," while in Hawaii, it is called the "purple plague." Various methods have been attempted in Hawaii to control *Miconia*, but none have proven very effective. As a result, teams of volunteers dedicated to protecting natural resources hike into the mountains and manually remove *Miconia* plants one by one.

[3] An organism whose genetic material has been changed using modern biotechnology, such as adding a gene from another species or altering its own genes.

Fig. 2.3 *Miconia*, an invasive species referred to as the "green cancer" in Tahiti. *Miconia* blocks light with its large leaves, preventing other plants from growing beneath it

Is It Advantageous for Plants to Be Taller Than Their Neighbors?

Many people wish they were a bit taller. Taller people tend to stand out and are often preferred. Is a taller plant preferred over shorter neighbors? Being tall makes it easier to access light and might offer other advantages too. For example, pollinators like bees, butterflies, or birds might be more likely to visit a taller plant first. Does this mean plants are constantly competing to grow taller?

But could there be drawbacks to being tall? In strong winds or heavy rain, a short plant with sturdy stems and roots may survive better than a tall, lanky one, which could topple over and become submerged. One fungus that parasitizes rice, *Gibberella fujikuroi*, secretes a toxin, which causes the rice plant to grow taller and become more prone to falling over. When the rice falls and becomes submerged in water, the fungus can feast on the leaves. Japanese farmers referred to such rice plants as "foolish seedlings," meaning they are tall but fail to produce a useful harvest. Additionally, when herbivores and insects attack from above, short plants may be targeted later than tall ones, giving them extra time to prepare defenses.

Then, perhaps, the ideal height for survival is neither too tall nor too short compared to neighboring plants. Supporting this idea, the chives in my home garden did not vary much in height. However, there may be alternative explanations. Could it be that they did not need to compete because they were all receiving enough light? Or perhaps, they were similar in height because they were genetically alike and, recognizing this, "chose" not to compete. In other words, either they had no advantage to gain in a competition, or their "kin recognition" prevented their competition.

I asked this question to Professor Yeonsook Choung from Kangwon National University, an expert in ecology, and she provided the following explanation. Crops like barley, chives, cabbage, and radish have similar seed sizes when planted, and the field conditions where they grow are also consistent, so their heights do not vary much. In contrast, wild plants produce seeds of varying sizes that fall into soils with diverse conditions. Therefore, it is not appropriate to infer the ecology of wild plants based on how crops grow. The height of plants varies greatly depending on environmental conditions, with light being the most important factor. When light is abundant, plants allocate fewer resources to growing taller, focusing instead on producing leaves for photosynthesis and roots for nutrient absorption. In natural conditions, among plants of the same age group that sprout and grow simultaneously, there are typically more short individuals and fewer tall ones. Tall individuals tend to have thinner, more delicate stems, while shorter ones develop sturdier structures. Survival advantages depend on the environment. For example, if deep shade suddenly sets in, taller plants are more likely to survive, while in windy highland areas, shorter plants are better suited to thrive. Thus, height is not a reliable predictor of survival. Instead, weight is a better indicator. In the absence of extreme environmental conditions, individuals that are heavier, moderately tall, and possess strong stems and roots are more likely to survive.

Forest managers and reforestation experts often use the principle of light-dependent plant growth. Initially, they plant trees densely to encourage rapid vertical growth as the trees compete for light. However, if the trees grow tall but develop weak trunks, they risk falling in strong winds. To prevent this, forest managers thin the trees once they reach a certain height, reducing competition and allowing the remaining trees to grow stronger and healthier.

A Walking Palm Tree That Moves to Get Light?

The palm tree *Socratea exorrhiza*, which lives in the tropical rainforests of Central and South America, is famous as the "walking tree" or the "walking palm." This tree develops stilt roots that grow from about 1 meter above the trunk and anchor into the ground (Fig. 2.4). According to those who claim that the tree walks, the dense rainforest offers little light, prompting the palm to move toward sunlight by extending stilt roots in the direction of sunlight, while allowing roots on the opposite side to die off, gradually shifting its position. Some say this palm tree can move several

Fig. 2.4 The stilt roots of *Socratea exorrhiza*, known as the "walking palm". These roots grow from about 1 meter above the trunk and anchor into the ground. This structure is highly effective in supporting the tree

meters in a year. This explanation of how the palm tree moves is appealing and has been featured in various media. But, does the walking palm tree really "walk" toward the light?

However, according to scientists who study *Socratea exorrhiza*, this palm tree does not move. While it is true that its stilt roots spread widely into the surrounding soil, the tree does not kill roots in a specific direction to enable movement. The nickname "walking palm" likely originates from the appearance of its stilt roots, which extend from the trunk above ground and resemble the legs of a walking animal. This visual similarity, combined with a seemingly plausible explanation, has led many to believe that this palm tree moves toward the light.

Then, what role do the stilt roots of *Socratea exorrhiza* serve? Stilt roots are highly effective at stabilizing the tree's structure. Unlike other trees that expend a lot of nutrients to develop thick trunks and deep underground roots for support, *Socratea exorrhiza* secures stability through its stilt roots, allowing it to allocate more nutrients toward the growth of its stem and leaves, enabling it to grow taller and faster. This tree does not "walk" on the ground to seek light; instead, it grows upward toward the sky. Through this strategy, the palm tree gains a competitive advantage for light among the densely packed plants in the tropical rainforest.

Competition for Space

Many people work hard with the belief that they should at least own a home. It seems natural to want to secure a space where one can rest comfortably and raise a family. However, the problem is that there is not enough good space available, so people must compete for it. Similarly, plants, like humans, are living beings with physical bodies, and they also require space to live. So, do plants also compete for space?

Indeed, they do so, endlessly. While trees compete by growing upward to claim well-lit space, grasses spread outward to cover the ground. The *Sequoia* trees in the forests of Redwood National Park in the United States, which can grow as tall as 100 meters, extend their trunks vertically and develop leaves high up (Fig. 2.5). If they are shorter and remain under other trees, their leaves will not be able to receive enough light for photosynthesis, which puts them at a major disadvantage for growth. These trees reach heights comparable to a 33-story building, similar to the Statue of Liberty or the Empire State Building. To support their weight as they reach upward, the *Sequoia* trees grow thick trunks. In essence, they strive to claim the sky.

Are there any drawbacks to such impressive height? The primary challenge for tall trees is transporting water and nutrients absorbed from the soil to the leaves at the very top. These trees have numerous thin tubes called xylem within their thick

Fig. 2.5 Plants competing for vertical space. In the forest, *Sequoia* trees compete for vertical growth in order to receive more light. (Photo by habitat_jaime, via iNaturalist. Licensed under CC BY 4.0)

Fig. 2.6 Reeds (left) and grass (right) spreading out horizontally to occupy space. Reeds extend underground rhizomes, while grass spreads stolons above the ground, from which stems and leaves are formed through clonal growth. Clonal growth is an effective way of occupying space

trunks to facilitate this movement, but lifting water against gravity to such heights is difficult. As a result, these trees cannot produce large leaves at the top, which would consume a lot of water. Although they can produce small leaves at the top, even these are not very useful for the tree's survival if they do not receive enough water to carry out active photosynthesis [65]. In other words, these trees have reached a height limit beyond which further growth offers no additional benefit.

However, some intriguing questions remain. Why did the species evolve to grow so tall? Was it because all other environmental conditions in their habitat were favorable, making light the limiting factor and thus driving them to such heights? Why did not these plants stop competing for height earlier? Were they be genetically programmed to continue competing for height? Now that scientists have sequenced the genes of the plant species, they may soon uncover compelling explanations for these questions.

While some plants compete by growing upward, others compete by spreading horizontally to occupy the ground. Many perennial herbaceous plants[4] create clones to secure more land. Plants like reeds, bamboo, grass, mugwort, and ferns spread widely by sending out roots or stems underground, engaging in clonal growth (Fig. 2.6). These plants wage a "guerrilla war" by expanding their roots or stems into any small gap they find, claiming the land as their own. Horizontal growth through cloning requires less material and energy compared to vertical growth, as the plants can obtain water and nutrients locally as they spread. However, it is not easy for these plants to penetrate areas already occupied by other plants. But when there is an ecological disturbance, such as a wildfire or tree logging, clonal plants can enter and grow vigorously. In fact, it is common to see ferns spreading rapidly and extensively through clonal growth in areas affected by wildfires. A clone refers to genetically identical individuals resulting from clonal growth, and no matter how widely they spread, they can all be considered a single organism. This is why, through genetic analysis, ferns over 1000 years old have been discovered. In the

[4] Herbaceous plants are plants with soft, nonwoody stems. In most cases, their aboveground parts die back in winter, while the roots survive and regrow in the spring.

end, plants striving to occupy space are not so different from us. It seems fair to say that all living beings are constantly seeking space to live and trying to expand it.

What happens when a large tree and small plants grow together in the same place? If the large tree develops many shallow roots that spread out and absorb most of the water and nutrients, it becomes difficult for smaller plants to grow nearby. However, if the large tree develops deep roots that reach far into the ground, it can actually help the smaller plants grow. When the deep roots draw water from the deeper soil, since the water is distributed throughout the entire plant body, the water concentration in the shallow roots increases. As a result, the roots near the dry surface have a higher water concentration compared to the surrounding soil, causing water to diffuse from the tree's roots back into the soil. This process allows smaller plants to utilize the water that the large tree has brought up from the deeper ground. While it may seem to us that these plants are coexisting in the same space, underground, they are positioned differently, with the large tree bringing resources that benefit the smaller plants.

Do Not Come Near Me: Allelopathy

As we have seen, plants compete with each other above ground to receive more light. But does such competition also occur underground? Yes, it does. Even though we cannot see it directly, plants compete underground for water and mineral nutrients (such as nitrogen, sulfur, phosphorus, potassium, calcium, and magnesium). Plants have to absorb water and mineral nutrients from the soil, and because these nutrients are not always in abundant supply and also because other plants need the same resources, competition is unavoidable. So, how do plants compete with each other underground?

When competing underground, plants often use chemical substances. Many types of chemicals secreted from the roots of plants are the weapons of this chemical warfare. Plants invest 20–40% of the substances produced through photosynthesis into making chemicals secreted from the roots [6], and many of these secreted substances are allelochemicals, which hinder the survival or growth of other plants. These allelochemicals secreted from plant roots cause a phenomenon known as allelopathy, preventing nearby plants from growing. When other plants are unable to grow, the water and mineral nutrients in the soil remain available, making it easier for the allelochemical-secreting plant to survive.

In laboratory experiments, the effects of allelochemicals were clearly observed. Research has shown that chemical substances in pine needles inhibit the germination and growth of various herbaceous plants, including perilla, horseweed, bittercress, and barnyard grass [61, 63]. In some cases, the chemical identities of the allelochemicals have been revealed. Phenolic compounds, such as tannins and coumaric acid, are known to have allelopathic effects. These compounds damage the roots of neighboring plants, thereby reducing their ability to absorb nutrients from the soil and ultimately inhibiting their growth. They also suppress fungi that form symbiotic relationships with plants [80]. Walnut trees release a volatile substance called juglone, which prevents the growth of other plant species nearby. Conversely, small plants like ferns can affect the growth of larger tree species, especially during the seedling stage, when their impact is more pronounced.

In laboratory experiments, the effects of allelopathy are clearly visible, but determining the significance of allelopathic substances in a natural forest setting is challenging. Even if a particular plant grows less successfully in one forest than in another, it is difficult to conclude that allelopathic substances secreted by other plants are the cause, as many alternative explanations are possible. It could be that the existing plants have already absorbed most of the water and nutrients, leaving less available for others. Or, it could be that materials like pine needles have accumulated and formed a thick layer, making it physically difficult for seeds to germinate and grow underneath. For instance, when plants flourish after a wildfire, one might argue that the fire incinerated allelopathic substances, enabling other plants to thrive. However, it may also be that the trees previously occupying the area were burned away, permitting more light to reach the ground, while nutrients in the ash supported new plant growth. Therefore, outside controlled laboratory conditions, confirming whether allelopathy plays a significant role for plants growing in forests remains challenging.

An Invasive Species Displacing Native Plants Through Allelopathy

Let us examine a case study on allelopathy. *Centaurea stoebe*, commonly known as spotted knapweed, is a small plant native to Europe. After its introduction to the United States, it spread rapidly, becoming a troublesome weed. Spotted knapweed, a perennial plant in the Asteraceae family, caused no significant issues in its native habitat. However, in the United States and Canada, it thrived, overtaking native plant species in pastures and grasslands. It grew densely even in barren and degraded environments, covering 4.5 million acres in Montana alone. This invasion reduced plant diversity, destroyed wildlife habitats, and severely disrupted livestock farming.

The first question scientists asked in this case was how this plant could suppress the growth of other plants and spread so widely. They hypothesized that spotted knapweed might release toxic substances from its leaves and roots that inhibit the growth of neighboring plants. To test this, they collected substances secreted from the roots of spotted knapweed and mixed them into plant culture media. As expected, the growth of other plants decreased. When activated carbon was added to the culture media to absorb and remove the substances secreted from spotted knapweed's roots, the inhibitory effect was reduced. This suggested that chemicals released from spotted knapweed's roots were responsible for suppressing other plants. Therefore, researchers concluded that spotted knapweed exhibits allelopathy. The next question was: What chemical substances are responsible for this allelopathic effect?

To identify the chemical responsible for the allelopathic effect, scientists collected substances secreted by spotted knapweed, chemically purified them, and tested each one separately in plant culture media to observe its impact on plant growth. Among the various substances secreted by the roots, they identified catechin as the one that consistently and effectively inhibited the growth of other plants. Catechin, a polyphenol, is known for its astringent taste in green tea and its antioxidant and medicinal properties. While this experiment suggested that catechin might be responsible for spotted knapweed's allelopathy, the conclusion was not definitive.

Some scholars argued that the results showing catechin inhibits plant growth were obtained in laboratory settings, and in natural environments, other chemical substances might play a more important role. Moreover, there were doubts about whether catechin was present in the soil at concentrations high enough to inhibit the growth of other plants. This controversy arose because different scientists measured different concentrations of catechin in the soil. There are several possible reasons for this. In some cases, catechin might have decomposed in the soil, or the efficiency of extracting catechin from the soil may have varied, or the amount of catechin secreted by the plant may have fluctuated due to environmental factors.

What kind of experiment would be needed to definitively confirm that catechin causes allelopathy? The research team led by Broz suggested genetically modifying spotted knapweed so that it cannot produce catechin. If this modified spotted knapweed fails to inhibit the growth of other plants, then it could be confirmed with confidence that the catechin secreted by spotted knapweed is the main substance responsible for its allelopathic effects, displacing native species [14].

There are still unresolved questions regarding the allelopathy of spotted knapweed, extending beyond just the role of catechin. For example, why did spotted knapweed not dominate in Europe? Several possibilities can be considered. First, plants in Europe may have coexisted with spotted knapweed for a long time and thus developed a tolerance to its allelopathic compounds, whereas North American plant species might have been more vulnerable to these compounds. Secondly, spotted knapweed may have produced more allelopathic compounds in North America as an adaptation to the new environment, potentially leading to the displacement of native plant species. Since the causes of ecological changes cannot be determined by studying a single species over a short period, further research is needed to answer this question definitively. However, as demonstrated in this example, when an invasive species rapidly displaces native species in a new environment, allelopathy is a likely factor.

Applications of Allelopathy

Research on allelopathy could improve agriculture, for example, by using allelopathic substances as herbicides to address issues associated with synthetic herbicides. Synthetic herbicides, which do not readily decompose in nature, can leave residues on crops, potentially leading to human absorption and associated health risks. Additionally, if residual herbicides spread from farmlands to surrounding environments, they can disrupt ecosystems. In contrast, allelopathic substances secreted by plants can selectively inhibit the growth of certain plants without affecting others and are more likely to decompose naturally, making them a potentially valuable alternative. If crops could be engineered to produce their own allelopathic substances, herbicide applications might be unnecessary, as weed growth would be inhibited. However, no allelopathic substances are currently widely used in agriculture. This is because their weed-inhibiting effects are much weaker than those of existing herbicides, and it remains uncertain whether these substances are genuinely safe for animals, humans, and the environment. The assumption that naturally produced chemicals are inherently safer than synthetic ones is flawed. Tens of thousands of plant-produced substances are already known, with likely many more undiscovered, and some are highly toxic to mammals or even carcinogenic. Many people mistakenly believe all natural substances are safe and all artificial ones are dangerous, but this is not a scientific perspective.

3 Parasitism in Plants

Parasitic Plants Siphon Nutrients from Host Plants

Most plants are autotrophic, meaning they produce their own energy and do not depend on other living things to obtain nutrients; they photosynthesize to generate their own energy and absorb water and nutrients from the soil. However, some plants rely on other plants for energy and nutrients and are called parasitic plants. Approximately 1% of flowering plants possess the ability to parasitize other plants. Initially, parasitic plants develop a structure called a *haustorium*, which attaches to the host plant. Next, the parasite connects its own pipe system, called vascular system, to that of the host (Fig. 3.1). Since vascular system is a network that transports water and nutrients, the fusion of the vascular systems of the parasite with that of the host enables the parasite to draw water and nutrients directly from the host. In this way, the parasitic plant extracts resources from its host much like a mosquito inserts its proboscis into a blood vessel to draw blood.

Parasitic plants that rely entirely on their host lack chloroplasts and do not carry out photosynthesis. *Hemiparasitic* plants, on the other hand, can photosynthesize on their own but still obtain some water and nutrients from the host plant. A commonly encountered hemiparasitic plant is European mistletoe (*Viscum album* var. *lutescens*).

The parasitic plant dodder (*Cuscuta japonica*) lacks leaves and roots and grows by wrapping its stem around the host plant's stem to absorb water and nutrients (Fig. 3.2). Among the crops particularly harmed by this plant is soybean. It is common to see a single dodder plant intertwining around the stems of several neighboring host plants, forming a clustered, almost communal structure. While dodder is a harmful parasitic plant that siphons nutrients from its host, research has shown that it can also act as a conduit for information between multiple host plants [50]. This study demonstrated that when one soybean plant is attacked by insects, other soybean plants connected to it through dodder also activate their defense responses.

Y. Lee, B. Y. Choi, *Social Lives of Plants*,
https://doi.org/10.1007/978-981-96-9837-0_3

Fig. 3.1 How the parasitic plant *Cuscuta* obtains sugars and nutrients from the host's phloem. The vascular bundle, which transports nutrients and water throughout the plant, serves as a conduit for this process. *Cuscuta* connects its haustoria to the host's vascular bundle, effectively stealing nutrients. (1) *Cuscuta*, (2) host plant, (3) *Cuscuta* stem, (4) host cells, (5) host phloem, (6) sugars and nutrients, (7) host epidermis, (8) *Cuscuta*'s haustorium growing toward and penetrating the host's phloem

Fig. 3.2 Image of a parasitic five-angled dodder (*Cuscuta pentagona*) entwining its host sea bindweed (*Calystegia soldanella*). The dodder is a parasitic plant made up solely of stems, with no leaves or roots. It attaches its vascular bundles to those of sea bindweed, almost as if they were grafted. (Grafting is the process of joining a part of one plant to another so that they grow as a single plant. In agriculture, this technique is used to propagate crops with desirable traits, such as seedless citrus fruits)

Defense compounds, such as jasmonic acid and substances produced by the attacked plant, travel through the dodder's vascular system to neighboring plants, triggering defense-related genes and prompting the production of protective substances in plants that were not directly attacked. As a result, these connected plants become better equipped to ward off insect invaders. One might imagine that dodder helps host plants survive since its own survival depends on the health of its hosts. However, researchers interpret this as a passive effect: the host plants' interconnected vascular systems through the dodder allow warning signals of insect attacks to spread to other hosts.

Another fascinating phenomenon is that dodder blooms only when its host plant blooms. Recent studies have revealed that the dodder genome lacks genes to control flowering time. Instead, proteins that induce flowering, produced by the host plant, are transferred to the dodder, prompting it to flower [117]. This synchronization of blooming periods likely provides a reproductive advantage for the parasitic plant. If dodder were to flower much earlier than its host, its growth period would be short, and it would not have sufficient resources to produce a large number of seeds. Conversely, if it were to bloom much later, the host plant would have already used up its nutrients in flowering and seed production, making it difficult for the dodder to secure the nutrients necessary for its own seed formation. Therefore, by timing its bloom to coincide with that of the host, dodder maximizes its seed production. It appears that to achieve this alignment in flowering time, dodder has undergone regressive evolution,[1] shedding genes related to its own flowering regulation.

Striga: A Parasitic Plant That Devastates Agriculture

Striga, a parasitic plant from the broomrape family (Fig. 3.3), causes severe agricultural damage across parts of Africa and Asia. This parasite primarily targets crops like maize, sorghum, sugarcane, and rice, stunting the growth of host plants and causing them to wither, which reduces crop productivity by about 40%. Sometimes, it even causes total crop failure. This parasitic plant affects approximately 300 million people in Africa, with subsistence farmers bearing the brunt of its devastating impact on their livelihoods.

How could *Striga* cause such great harm? One *Striga* plant can release over 100,000 tiny, dust-like black seeds. These seeds remain dormant in the soil, waiting for chemical signals that indicate the presence of a host plant nearby. This waiting period can last for over 10 years, but once they detect the host plant, they germinate and attach to it. *Striga* then connects its vascular system to that of the host, siphoning water and nutrients.

[1] Regressive evolution refers to the process by which organisms gradually lose traits that are no longer beneficial. For example, modern humans share a common ancestor with other primates. Over time, the tail—once a functional structure—underwent regressive evolution and now exists as a vestigial organ called the coccyx.

Fig. 3.3 *Striga*, a parasitic plant that causes serious damage to African agriculture. (Photo by Tropentag, via Flickr. Licensed under CC BY-ND 2.0)

The chemical signal indicating a host plant's presence to *Striga* is a compound called strigolactone, which plants release to establish symbiosis with fungi. For the plant, it is like opening the door to welcome a friend, only for an enemy to enter instead. Around 80–90% of terrestrial plants form symbiotic relationships with fungi, which develop hyphae to absorb water and nutrients like phosphates and provide them to the plant. Without the fungi's support, plants struggle to thrive in poor soils, so they release strigolactones to attract these beneficial fungi. *Striga*, however, uses this compound to detect the presence of a host.

Strigolactone was first identified by American scientists in 1966 [23]. The team isolated a potent germination stimulant for witchweed (*Striga lutea*) seeds from the root exudates of the host plant, cotton. This stimulant was highly effective in inducing *Striga* seed germination even at very low concentrations. It was named strigol due to its effect on *Striga* germination. Later, other related compounds with similar germination-stimulating effects were identified. These compounds are collectively called strigolactones, named for the presence of a lactone ring, which is essential for their function as germination stimulants.

When soil nutrients are sufficient, plants no longer need the fungal assistance, and they reduce strigolactone release. With reduced strigolactone levels, *Striga* cannot detect the presence of a host and does not germinate. Consequently, *Striga* is not much of a problem in wealthier countries that use ample fertilizers but is a serious issue in regions like Africa, where soils are poor and farmers often cannot afford fertilizers.

After hearing these stories at a conference, I wanted to study this plant and started looking into how to get *Striga* seeds. However, *Striga* is such a dangerous plant that its import into Korea is prohibited, and trying to do so secretly could lead to legal trouble. Instead, I thought of obtaining strigolactone to identify the plant gene encoding the transporter protein that releases strigolactone into the soil. However, strigolactone acts in minuscule amounts, and plants produce very little of it. Synthesizing it is also extremely difficult, which makes it expensive even into the early twenty-first century. I suppose these are the reasons why not many scientists can study how to control this parasitic plant.

Research to Reduce the Damage from *Striga*

Protecting African crops from *Striga* would be immensely beneficial for the people of Africa. To tackle this issue, scientists have devised various strategies and continue to study potential solutions. One approach considered was to eliminate *Striga* seeds, which are scattered like black dust across fields and impossible to remove individually. A simple idea leverages *Striga*'s dependence on strigolactone for germination. By applying strigolactone to fields before planting crops, the *Striga* seeds would be induced to germinate. Once germinated, *Striga* seeds cannot return to dormancy, and without a host, they would die due to lack of water and nutrients. As a parasitic plant, *Striga* cannot survive without a host. This would, in theory, leave the fields safe for subsequent crop planting. Scientists have attempted to find structurally similar, less costly compounds that would trick *Striga* into germinating, and some positive results are just starting to come out. However, no such compound has yet been employed in the field, and the current strategy is to plant crop varieties that are more resistant to *Striga*.

Recently, a research team led by Professor Al-Babili at King Abdullah University of Science and Technology in Saudi Arabia discovered three relatively cheap compounds similar enough to strigolactone to make *Striga* germinate [69]. When tested in pots, these compounds successfully triggered *Striga* seed germination, and they also showed good results in field trials in Burkina Faso, West Africa. The compounds were applied to fields twice after rainfall, and crops were planted 10 days later. In fields treated with these compounds, *Striga* infestations dropped by up to 65% in millet fields and 55% in sorghum fields compared to fields treated with just water. However, *Striga* was not completely eliminated because its seeds do not all germinate at once but rather at different times throughout the year. In theory, applying the compounds multiple times to an empty field could eventually make all the *Striga* seeds germinate and die. But in reality, poor farmers cannot afford to leave their fields empty just to remove *Striga*. If this method continues to be explored, and more effective, affordable compounds are made available to African farmers, it could be a helpful solution for Africa's food security problems.

The Host Plant's Strategy to Defend Against Parasitic Plants

So, does a plant just let a parasite take its nutrients? Are there no plants that fight back against parasites? Yes, some plants do have ways to defend themselves. Some "hide" to avoid being detected by parasites, while others produce chemicals that harm the parasites or stop them from attaching.

A variety of sorghum resistant to *Striga* uses the hiding method. It subtly alters the stereochemistry of the strigolactone it secretes from its roots, making it

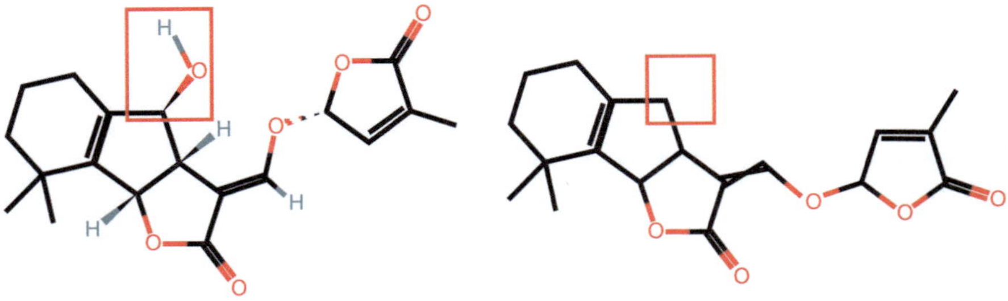

Fig. 3.4 The chemical structures of orobanchol and 5-deoxystrigol. The structures of orobanchol (left) and 5-deoxystrigol (right) differ only in stereochemistry. While *Striga* seeds recognize 5-deoxystrigol as a host-released compound and germinate in response, they do not recognize orobanchol. Beneficial fungi, however, can recognize both orobanchol and 5-deoxystrigol, initiating symbiosis with the plant roots. Sorghum variety resistant to *Striga* releases more orobanchol than 5-deoxystrigol, thereby reducing the germination and parasitism of *Striga* seeds

unrecognizable to *Striga* (Fig. 3.4). When the sorghum releases this modified strigolactone, symbiotic fungi can still detect the plant, but *Striga* cannot, so they do not germinate. From the sorghum's point of view, it has successfully hidden from *Striga*.

Certain sunflower species and wild sorghum use the chemical defense strategy. The sunflower releases a toxic chemical called coumarin from its roots, which inhibits the growth of parasitic plants. Wild sorghum disrupts the attachment process of parasitic plants by releasing a compound from its roots that inhibits the formation of haustoria, structures essential for the parasite to attach.

Some details of the mechanisms by which plants recognize and mount defenses against parasitic invasion have been revealed. The host plant's cell membrane contains receptors that detect the presence of a parasitic plant. When these receptors recognize substances secreted by the parasite, they trigger a hypersensitive response. This response produces high levels of hydrogen peroxide, which kills cells along the path of the parasite's invasion. Hydrogen peroxide also induces the synthesis and release of defense hormones, strengthens the cell wall, and promotes the production and release of toxic compounds. Together, these actions prevent the parasite from forming straw-like structures that connect its haustoria to the plant's vascular system to siphon nutrients. Overall, this mechanism resembles the plant's response to pathogens like bacteria and fungi (Fig. 3.5).

In Part I, we explored how plants interact with one another and survive within their ecosystems. Since resources—such as light, inorganic nutrients, space, and pollinating insects—are not unlimited, fierce competition arises among plants to obtain them. Beyond mere competition, some parasitic plants survive by seizing what other plants have acquired. However, competition alone does not define the social interactions between plants. It has been discovered relatively recently that most plants in nature are physically interconnected through their roots, allowing them to

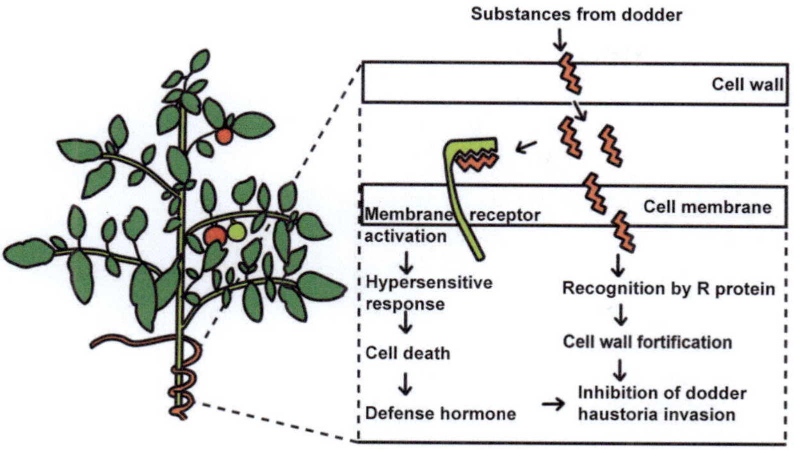

Fig. 3.5 Tomato's defense mechanism against parasitic dodder. The tomato plant has receptors in its cell membrane that recognize substances from dodder and activate upon detection. These activated receptors trigger a hypersensitive response, causing cell death in the area where the dodder attempts to attach, and activate the production of defense hormones. Additionally, the tomato plant has R (resistance) proteins inside the cell that recognize substances from dodder and initiate the reinforcement of the cell wall. Together, these reactions prevent dodder from penetrating

share information and nutrients. Furthermore, when pathogenic microorganisms or pests invade, plants often cooperate to fend off these threats.

Thus, both *competition* and *cooperation* are essential for the survival of plants, and *plants live in constant social interactions with other plants*. In fact, no single organism can live entirely isolated; nature does not allow anything to live in complete solitude. Furthermore, a plant's strategy of social interactions must change in response to changes in the environment and the strategies of neighboring plants: it has to evolve to remain effective. The plants we see today are remarkable life forms that have adapted to environments where social interaction is inevitable. They have found their niche and survived.

Part II

Plants and Microbes

Plants coexist with numerous microorganisms on and around their bodies. Some of these microorganisms reside on the plant's surface, while others inhabit empty spaces within leaves or roots, or even live inside plant cells. Many also thrive in the air, soil, and water surrounding the plant. While bacteria and fungi are the most commonly known microorganisms associated with plants, others include protozoa,[1] nematodes,[2] viruses, and microalgae. Collectively, these microorganisms are referred to as the plant microbiome (Fig. 1). Among the plant microbiome, some

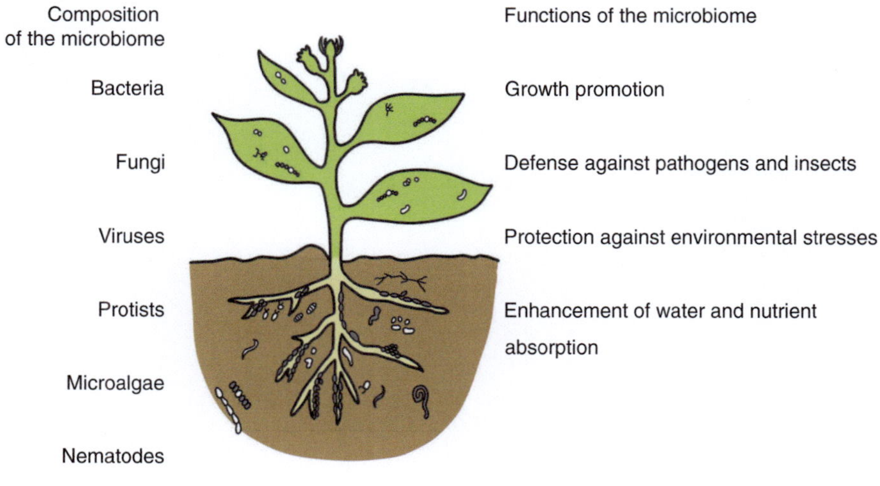

Fig. 1 Composition and roles of the plant microbiome. Plants coexist with microorganisms such as bacteria, fungi, protozoa, nematodes, viruses, and microalgae. These microorganisms, collectively referred to as the "plant microbiome," generally support plant growth, nutrient absorption, resistance to pests and diseases, and environmental adaptation

[1] Protozoa refer to various types of single-celled organisms that have a nucleus in their cells but do not belong to the animal, plant, or fungal kingdoms.

[2] Nematodes are unsegmented roundworms found in a wide range of environments. Many species are parasites of plants, animals, and humans.

microorganisms benefit the plant, others have no noticeable effect, and some are harmful—much like the beneficial and harmful microorganisms found in the human gut.

Research on microorganisms interacting with plants has traditionally focused on how to protect crops from pathogens that cause diseases. Plant pathogens are adept at infecting crops, and when crops become diseased, agricultural yields can suffer devastating losses. In contrast, beneficial microorganisms associated with plants received relatively little attention until the early twenty-first century. However, recent findings highlighting the microbiome's critical role in plant survival have spurred rapid advancements in this area of research. Although their roles vary, plant microbiomes generally support nutrient absorption, pest and disease resistance, and environmental adaptation [131].

In this part, we will first explore the symbiotic relationships in which plants and microorganisms mutually assist each other for coexistence. Following this, we will examine the intense battles between plants and pathogenic microorganisms as they struggle for survival.

Plants in Symbiosis with Microorganisms

4

The Origins of Plant-Microbe Symbiosis

When did plants begin their symbiotic relationships with microorganisms? It is hypothesized that plant-microbe symbiosis began around 450 million years ago, when fungi known as endomycorrhizal fungi[1] formed mutually beneficial interactions with early terrestrial plants [26]. At the time when early terrestrial plants evolved from aquatic photosynthetic organisms to adapt to life on land, nutrients were scarce, making symbiosis with nutrient-supplying microorganisms crucial for their survival. The symbiosis between plants and nitrogen-fixing bacteria emerged later, approximately 100 million years ago, and developed independently multiple times. For microorganisms, plants provided a convenient and safe habitat, while plants benefited by internalizing them, as this made it easier to manage and utilize microorganisms effectively.

Dr. Delaux from France argues that symbiosis with microorganisms played a pivotal role in enabling plants to adapt to terrestrial environments and thrive extensively [104]. He provides four key pieces of evidence to support this claim. First, the oldest terrestrial plant fossils, dating back 407 million years, have traces of fungi as endosymbionts.[2] Second, symbiosis with arbuscular mycorrhizal fungi is not limited to flowering plants but is found across all terrestrial plants. Third, when examining the genes required for symbiosis, similar genes are found not only in flowering plants but also in simpler terrestrial plants such as ferns, liverworts, and mosses. The presence of these symbiosis-related genes in diverse terrestrial plants suggests they were inherited from the common ancestor of early terrestrial plants. Finally, when endomycorrhizal fungi begin to establish symbiosis with plants, they induce changes

[1] Endomycorrhizal fungi form a symbiotic relationship with plant roots, where fungal hyphae penetrate plant cells to establish mutualistic interactions.

[2] An endosymbiont is an organism that lives inside the cells or body of another organism in a mutualistic relationship.

Y. Lee, B. Y. Choi, *Social Lives of Plants*,
https://doi.org/10.1007/978-981-96-9837-0_4

in the expression of various plant genes, and these changes are remarkably similar in flowering plants and liverworts.

While symbiosis with microorganisms offers significant benefits to plants, it also comes with drawbacks. Pathogenic microorganisms can exploit the pathways plants create for their symbiotic partners. For instance, to initiate symbiosis, plants secrete enzymes that soften the cell wall by breaking it down, allowing microorganisms to enter. However, this softened cell wall also gives pathogens an opportunity to invade. It is like opening a door to welcome a friend, only to have a thief slip in as well. To survive, plants have developed strategies to distinguish friends from foes. In response, pathogens have evolved ways to overcome these defenses. Over hundreds of millions of years on land, the dynamic interactions among plants, symbiotic microorganisms, and pathogens have led to a wide array of social relationships in ecosystems.

Microorganisms Supporting Plant Growth

The microbiome generally benefits plant survival, particularly by playing a crucial role in nutrient absorption. Many microorganisms break down complex substances in the soil into simpler forms that are easier for plants to absorb. For example, some microorganisms decompose organic matter[3] originating from living organisms into ammonia, while nitrifying bacteria convert ammonia into nitrates, making these nutrients more accessible to plants. Nitrogen-fixing bacteria and mycorrhizal fungi, which we will explore further in this chapter, assist plants in nutrient uptake. Additionally, microorganisms boost a plant's stress tolerance by producing substances such as ROS-detoxifying enzymes and plant hormones. Furthermore, they enhance plant defense against pathogens by directly competing with harmful microorganisms, preventing their overgrowth, or by synthesizing and secreting antibiotics that kill pathogenic bacteria and fungi.

Different cultivars of soybean and tomato plants show varying levels of resistance to diseases. Studies suggest that cultivars resistant to diseases have a unique microbiome compared to those that are more vulnerable [70, 83]. But could the microbiome actually affect how plants resist diseases? A research team led by Professor Jihyun F. Kim at Yonsei University in South Korea made an important discovery about this. They studied a tomato cultivar called Hawaii 7996, which is resistant to bacterial wilt disease, and another cultivar called Moneymaker, which is susceptible to the disease but produces a higher yield. The disease is caused by the soil bacterium *Ralstonia solanacearum*. The team found that Hawaii 7996's resistance comes from the unique microbiome in the soil around its roots. This microbiome is rich in a beneficial bacterium called TRM1. What is even more exciting is that this protective effect can be transferred. When soil from Hawaii 7996 plants is mixed with soil from Moneymaker plants, the Moneymaker plants also become

[3] Organic matter refers to carbon-containing compounds, including those naturally produced by living organisms and those synthesized artificially in laboratories.

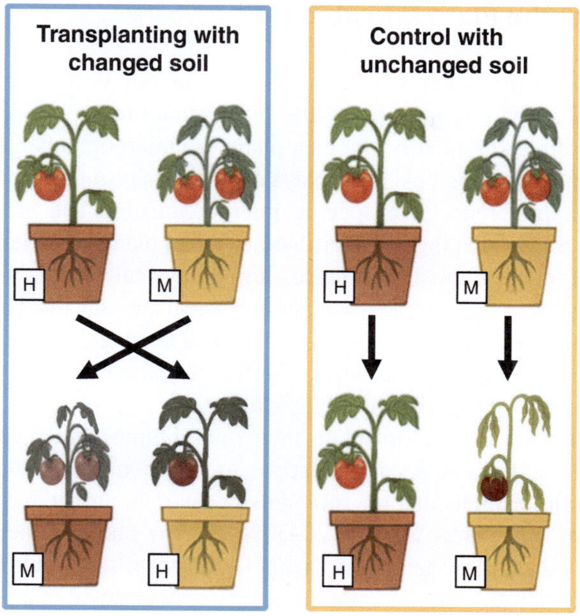

Fig. 4.1 Transplanting microbes can enhance disease resistance. When a disease-susceptible Moneymaker tomato cultivar (M) was grown in soil previously used for the bacterial wilt-resistant Hawaii cultivar (H), and the pathogen was introduced, disease progression in Moneymaker was suppressed—the disease developed more slowly and with milder symptoms. Conversely, when the Hawaii cultivar was grown in soil previously used for Moneymaker, the disease progressed slightly faster

more resistant to bacterial wilt (Fig. 4.1). This occurs because the microbiome from Hawaii 7996, including TRM1, suppresses the harmful pathogen in Moneymaker's soil. This finding offers a natural, sustainable strategy to reduce crop disease and enhance productivity on farms.

The soil microbiome in farmland that has been cultivated for a long time is very different from that of naturally undisturbed land. In agriculture, the practice of densely planting a single crop reduces the diversity and stability of soil microorganisms. As a result, the number and types of microbes that protect plants diminish, making crops more vulnerable to diseases. It has been suggested that, just as people take probiotics to improve their gut health, introducing beneficial microorganisms into farmland could enhance crop health and farming success. However, the effectiveness of this approach varies widely depending on environmental conditions and the microorganisms already present in the soil. The problem is even greater when only one type of microorganism is added to the soil. To solve this, a better approach might be to introduce a mix of diverse microorganisms. But what combination works best for farming? Many researchers are trying to answer this question. Success in this area could lead to farming methods that use less pesticide in the future.

Formation of the Plant Microbiome

The plant microbiome is not a random assembly of microorganisms. Then, what factors determine the composition of the plant microbiome? It is shaped by the interactions of three key factors: the host plant, the microorganisms, and the environment. Among these, the host plant plays a particularly significant role, as different plants have preferences for specific microorganisms. The environment also plays a crucial role in shaping the microbiome. When plants experience stress, they often alter their microbiome. For instance, during physical stresses such as drought, high temperatures, or low temperatures, or biological stresses like pathogen or pest attacks, the plant microbiome undergoes substantial changes. In other words, when the physical or ecological environment shifts, plants adapt by modifying their microbiome to better cope with the new conditions.

How do plants attract and maintain their microbiome? Much like we provide good food and comfortable accommodations to make our friend stay long, plants offer nutrients and habitats that microorganisms favor, to attract and maintain their desired microbiome. Plants allocate 20–30% of their photosynthetic products to secreting substances from their roots, primarily to establish symbiosis with microorganisms. If symbiosis with microorganisms were not critical to a plant's survival, it is unlikely that plants would invest such a significant portion of their resources into root exudates. In other words, generating such especially beneficial conditions around their roots likely gave plants an evolutionary edge over plants with fewer and less enriched exudates.

The soil surrounding plant roots is teeming with microorganisms, with estimates suggesting up to 100 billion microorganisms per gram of root [11]. Plants create a highly complex world within the soil, akin to another universe. By carefully selecting and managing the microbiome in the soil, as well as on their bodies, and in their surroundings, plants cultivate an environment that supports their growth. How did scientists uncover these fascinating insights? Let us now delve into the experiments that revealed these findings.

In 2019, a collaborative study by Dr. Osbourn's team in the United Kingdom and Dr. Bai's team in China revealed that the plant *Arabidopsis thaliana* synthesizes and secretes triterpenoids from its roots shaping a unique root-associated microbiome specific to the plant [53]. Triterpenoids are compounds commonly produced by trees and include squalene, steroids, and saponins. The genome of *Arabidopsis thaliana* contains numerous genes responsible for synthesizing triterpenoids, with four of these genes clustered on a single chromosome and expressed in the roots. The clustering of genes with similar functions on a chromosome often results from strong evolutionary pressure,[4] indicating that these genes were crucial for the plant's survival. Based on this information, the research team suspected that these triterpenoid-synthesizing genes have played a critical role in the survival of *Arabidopsis*

[4] Evolutionary pressure refers to environmental factors that affect an organism's ability to survive and reproduce. These pressures drive natural selection by reducing traits that are disadvantageous and promoting those that are beneficial.

thaliana over time. Since triterpenoids were already known to influence microbial proliferation, they hypothesized that "*Arabidopsis thaliana* uses these genes to shape its root-associated microbiome."

To test this hypothesis, researchers compared the microbiomes of mutant *Arabidopsis thaliana* plants, deficient in triterpenoid synthesis due to genetic deletions, with those of wild-type plants. They reported three key findings: (1) The microbiome of the mutants differed significantly from the wild type; (2) when 19 root-associated bacterial strains were directly treated, triterpenoids promoted the growth of some strains while inhibiting others; (3) mutants had a lower proportion of bacteria whose growth was enhanced by triterpenoids compared to the wild-type plants. The authors concluded that "*Arabidopsis thaliana* synthesizes unique triterpenoids to establish a specialized root microbiome" and speculated that plants in general may produce and secrete specific compounds to attract and maintain essential microbiomes.

Triterpenoids are not the only class of compounds that plants have been found to secrete to shape their root microbiome. Coumarins, phenylpropanoid-derived secondary metabolites[5] with a sweet yet slightly bitter taste, are abundant in kudzu roots and influence bacterial communities associated with *Arabidopsis thaliana* roots [124]. This study found that coumarins inhibit the growth of soilborne fungal pathogens like *Fusarium oxysporum* and *Verticillium dahliae*, which cause fusarium wilt and verticillium wilt, respectively. However, these compounds do not affect beneficial bacteria such as *Pseudomonas P417* and *Pseudomonas capeferrum*, which support plant growth and defense. By secreting coumarins, plants promote the proliferation of beneficial bacteria near their roots. In turn, these bacteria influence the plant's secretion of compounds; for instance, *Pseudomonas P417* stimulates *Arabidopsis thaliana* roots to secrete more coumarins, fostering an environment favorable for its survival while protecting the plant from fungal pathogens.

The two cases discussed above focus on the specific chemicals that plants secrete to shape and manage their microbiome, whereas the next study shifts attention to identifying the plant protein responsible for this function. Indica rice[6] outperforms Japonica rice[7] in nitrogen-deficient environments, largely due to its association with a greater number of nitrogen-metabolizing microorganisms [145]. A 2019 study by Professor Bai's laboratory in China investigated how these rice varieties attract distinct microbiomes by focusing on the nitrate transporter gene *NRT1.1B*, a key genetic factor in nitrogen use efficiency. The *NRT1.1B* gene in Indica rice differs

[5] Secondary metabolites refer to organic compounds synthesized by living organisms that are not directly involved in growth, development, or reproduction. Many secondary metabolites function as toxic substances that play a defensive role.

[6] A rice variety primarily cultivated in tropical regions with high temperatures and abundant rainfall throughout the year. It has a high amylose content and a low amylopectin content in its starch, resulting in cooked rice that is less glossy and less sticky. The grains are longer compared to Japonica rice.

[7] A rice variety cultivated in temperate regions, characterized by a low amylose content and a high amylopectin content in its starch. As a result, cooked rice has a glossy appearance and a sticky texture. The grains are rounder and shorter compared to Indica rice.

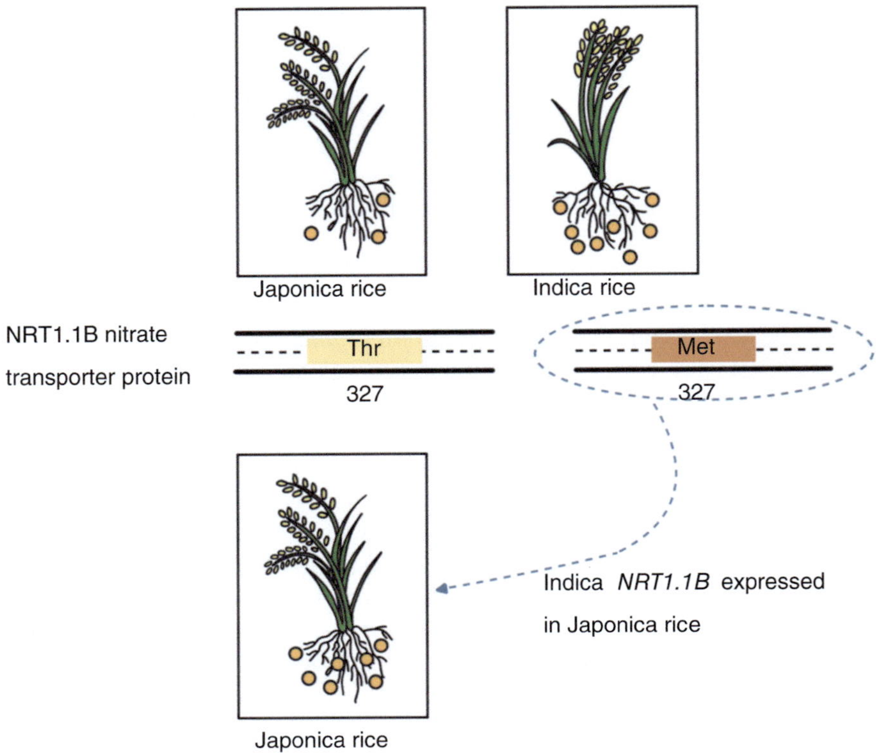

Fig. 4.2 A plant nitrate transporter protein alters the composition of root-associated microbiomes. Indica rice attracts more nitrogen-metabolizing microorganisms than japonica rice, enabling it to thrive even in nitrogen-deficient conditions. The NRT1.1B nitrate transporter protein in indica and japonica rice differs by a single amino acid at position 327. When the *NRT1.1B* gene from indica rice was expressed in japonica rice, the number of nitrogen-metabolizing microorganisms in the surrounding area increased. This result suggests that the NRT1.1B nitrate transporter protein plays a critical role in shaping the composition of microbiomes in rice plant

from that in Japonica rice by a single nucleotide,[8] leading to a single amino acid change. To test the role of this genetic difference, researchers artificially expressed the *NRT1.1B* gene from Indica rice in a Japonica rice cultivar and observed its impact on the microbiome (Fig. 4.2). The experiment revealed changes in seven microbial taxa, six of which were specifically associated with Indica rice. These findings suggest that the minor genetic variation in the *NRT1.1B* gene is crucial for attracting different microorganisms to Indica and Japonica rice. However, the researchers could not determine how *NRT1.1B*, a nitrate transporter, influences the

[8]A nucleotide is the basic building block of DNA, consisting of a sugar, a phosphate group, and one of four nitrogenous bases: adenine (A), guanine (G), cytosine (C), or thymine (T). The sequence of these bases within the nucleotides determines the amino acid sequence of the protein encoded by the gene.

microbiome's composition. They speculated that when Indica rice struggles to absorb nitrogen, NRT1.1B activates, altering root exudates to attract nitrogen-metabolizing microbes.

The research team studied how their findings could be used in agriculture by isolating more than a thousand bacterial strains from the roots of Indica and Japonica rice. They then created synthetic microbiomes using these bacteria. When they applied the Indica rice microbiome to plants grown with only large organic nitrogen molecules as the nitrogen source—without ammonium or nitrate—the plants grew better than those treated with the Japonica rice microbiome. This improvement happened because bacteria in the Indica microbiome converted large organic nitrogen compounds into ammonium and nitrate, which are easier for rice roots to absorb. Using these synthetic microbiomes could reduce the need for nitrogen fertilizers while still supporting healthy plant growth. This approach may lower fertilizer-related pollution and make farming more sustainable, because the generation of nitrate and ammonium by the root-associated synthetic microbiome would occur gradually and in close proximity to the roots, thereby minimizing leaching.

Symbiosis Between Plants and Fungi: Formation of Mycorrhizae

Let us now explore the symbiosis between plants and fungi, a relationship that began 450 million years ago when plants first expanded their habitat from water to land. This partnership continues to play a critical role in the survival of terrestrial plants, especially in nutrient-deficient environments. Today, about 80% of vascular plants form mycorrhizae to partner with fungi.

Mycorrhizae are categorized into two main types: ectomycorrhizae, where fungal hyphae form a sheath-like structure covering the surface of plant roots, and endomycorrhizae, where the hyphae penetrate plant cells (Fig. 4.3). The symbiosis between plants and endomycorrhizal fungi forms an extremely close relationship, extending even into the interior of plant cells.

How is such a close relationship formed? The first step in the formation of mycorrhizae involves the plant expressing its "intent" to establish symbiosis by releasing signaling molecules from its roots. Identifying the nature of these signaling molecules was challenging, leading scientists to hypothesize that they were likely highly unstable and present in very low concentrations in root exudates. To identify these molecules, researchers investigated whether any compounds secreted by plant roots could promote the growth of fungal hyphae. This approach was based on the understanding that the successful growth of fungal hyphae near plant roots is a prerequisite for establishing a symbiotic relationship. Consequently, the signaling molecules facilitating symbiosis were expected to enhance fungal hyphae growth.

Scientists separated root exudates into lipid-soluble and water-soluble components and further fractionated the lipid-soluble ones using various solvents. Each fraction was tested on fungi to observe hyphal growth. This process revealed that the fraction extracted with the organic solvent ethyl acetate promoted fungal hyphal

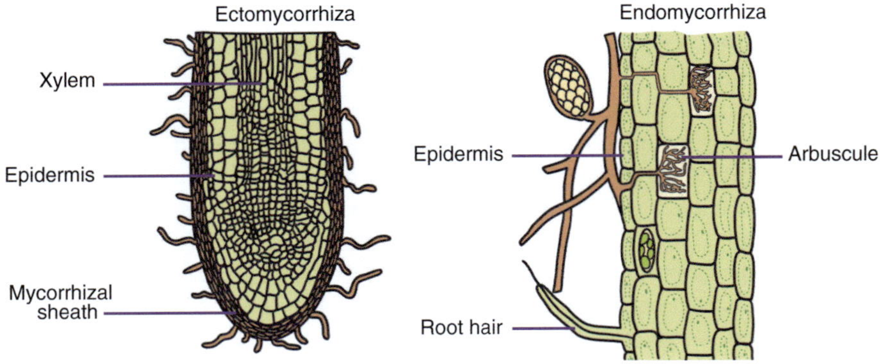

Fig. 4.3 Structures of ectomycorrhizae and endomycorrhizae. Ectomycorrhizae are characterized by fungal hyphae forming a sheath-like structure that covers the outer surface of plant roots. In arbuscular mycorrhizae, a type of endomycorrhiza, fungal hyphae penetrate plant cells and form highly branched, surface-enlarging structures called arbuscules. These arbuscules serve as the sites for material exchange between the plant and the mycorrhizal fungi

growth. Later, Japanese scientists hydroponically cultivated *Lotus japonicus*, a wild legume, under phosphorus-deficient conditions. They isolated ethyl acetate-soluble compounds from the plant's root exudates and, through further purification and testing, identified a compound called strigolactone, which promotes the growth of symbiotic fungal hyphae [2].

These biochemical experiments were followed by genetic and physiological experiments showing that strigolactones play a key role in plant-fungi symbiosis. In diverse land plant lineages—including dicots, monocots, bryophytes, and liverworts—disrupting strigolactone biosynthesis impairs symbiosis with endomycorrhizal fungi. However, applying strigolactones externally restores symbiosis even in these cases [66].

Strigolactone is a compound already well-known to plant scientists as a signaling molecule. Parasitic plants like *Striga*, a serious problem in Africa, sense it to locate their host plants. This means strigolactone serves as a signaling molecule in two contrasting ways: it supports symbiosis with beneficial mycorrhizal fungi but also enables parasitism by harmful plants. As mentioned earlier, while plants release strigolactone to attract helpful partners, "thieves" like parasitic plants detect this signal and exploit it, acting like uninvited guests crashing a party.

What happens after a plant secretes strigolactone? First, fungal hyphae sense the strigolactone released by the plant roots and start growing actively toward the plant. As the fungi get closer, they release signaling molecules that prompt changes in the plant's gene expression. Finally, when the fungal hyphae come into contact with the surface of the plant root, the plant's cytoplasm condenses, creating pathways for the fungi to grow and enter. The fungal hyphae then grow actively through these pathways into the plant cells, where they form highly branched structures called arbuscules (Figs. 4.3 and 4.4), establishing a stable symbiotic relationship.

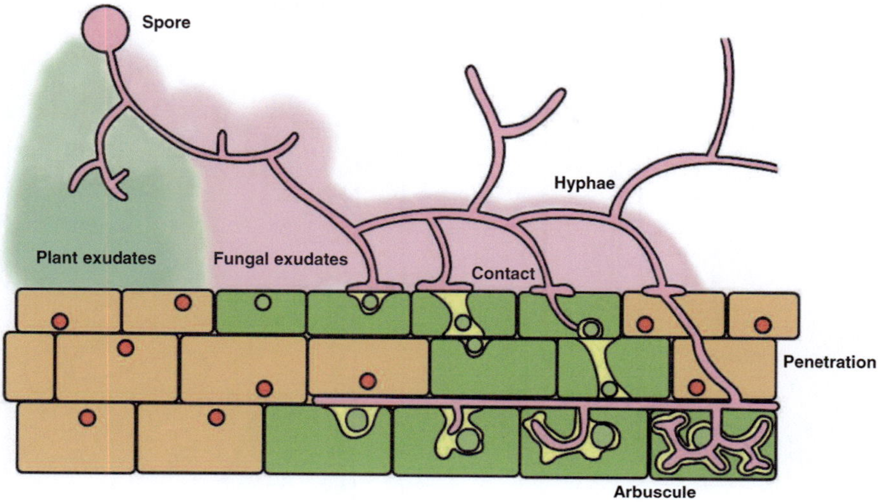

Fig. 4.4 Diagram of the formation process of endomycorrhizae. When fungal hyphae detect strigolactones secreted by the plant, they grow actively and move toward the plant. Upon contact between the growing hyphae and the surface of the plant root, the plant cell's cytoplasm condenses, forming a pathway for the fungus to enter and grow. The hyphae then grow along this pathway into the plant cells, where they develop highly branched arbuscules. These arbuscules, with their extensive surface area, facilitate efficient nutrient exchange between the plant and the fungus

Not all plants form symbiotic relationships with mycorrhizal fungi. Parasitic plants, such as *Striga*, and carnivorous plants, like the Venus flytrap, do not establish symbiosis with mycorrhizal fungi. The mechanisms that parasitic plants use to penetrate the vascular tissues of their hosts, or the structures carnivorous plants employ to digest insects, are highly specialized and resource-intensive. These plants have evolved to focus on one unique strategy, creating precise mechanisms for interacting with a limited number of organisms. In contrast, many other plants have developed multiple approaches, forming relationships with a diverse range of surrounding life forms. In summary, plants exhibit a wide variety of strategies, from specific interactions to broad relationships with their neighbors.

The Give-and-Take of Plant-Fungi Symbiosis

Symbiosis is a mutually beneficial relationship in which one organism provides what it can easily produce and receives what it lacks from its partner. So, what is easy for plants to produce? Plants can efficiently produce sugars and lipids. These compounds are composed of carbon, hydrogen, and oxygen, which plants can obtain by absorbing carbon dioxide from the air and water from the soil. On the other hand, synthesizing compounds like proteins and nucleic acids requires nutrients such as nitrogen (N), phosphorus (P), and sulfur (S), which are not always

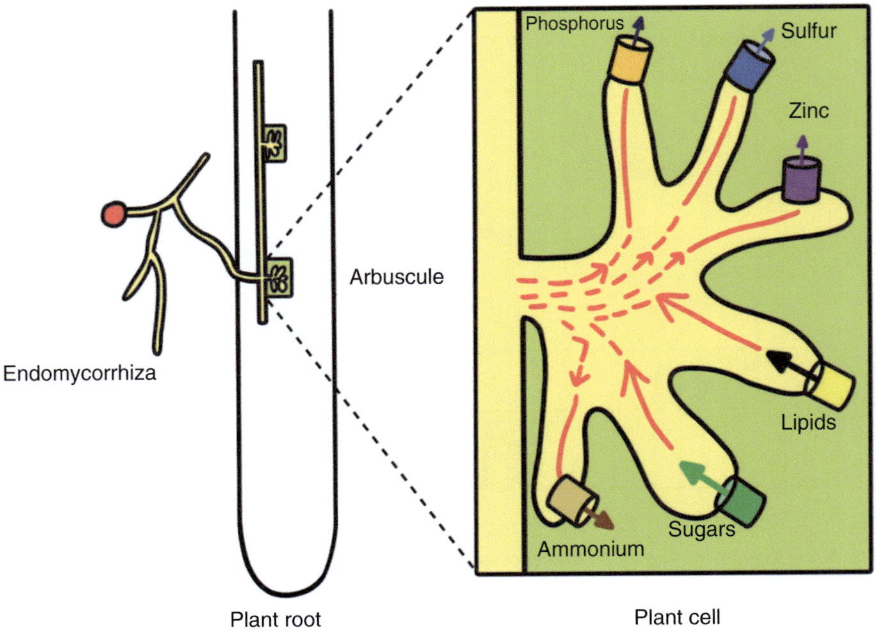

Fig. 4.5 Substances exchanged between symbiotic plants and mycorrhizal fungi. Plant cells supply sugars and lipids to mycorrhizal fungi, while the fungi provide the plant cells with inorganic nitrogen sources such as ammonium and nitrate, as well as other inorganic nutrients like zinc, sulfur, and phosphate. Note that the interface between the plant and the fungal hypha–represented here by a single line–consists of two juxtaposed relatively impermeable membranes of both. Consequently, the exchange of materials requires various proteinaceous transporters embedded in each membrane

readily available in the environment. Iron (Fe) is another crucial mineral for plant growth, but it is difficult for plants to acquire. In a symbiotic relationship with mycorrhizal fungi, plants provide sugars and lipids, which are easy for them to produce, in exchange for phosphorus, nitrogen compounds, sulfur, iron, and other essential minerals from the fungi.

At the interface where plant cells and symbiotic microorganisms meet, various transporter proteins facilitate the exchange of resources (Fig. 4.5). These proteins transfer sugars and lipids from the plant to the microorganisms and inorganic nutrients from the microorganisms to the plant. Among the plant's phosphate transporters, the PHT1 family plays a critical role in acquiring phosphate from endomycorrhizal fungi. In *Medicago truncatula* mutants missing PHT1 genes, arbuscule formation stops halfway, preventing further fungal growth within the plant tissues [55]. This may be because, without phosphate transport from the fungi to the plant, the plant fails to recognize the presence of beneficial fungi. As a result, it stops supplying sugars and lipids, disrupting arbuscule development.

It has long been known that plants supply sugars to their symbiotic fungi, but the discovery that they also provide lipids emerged much later. In 2010, researchers established that the ABC transporters STR1 and STR2 in *Medicago truncatula* are essential for the symbiosis between host plants and mycorrhizal fungi [144]. However, their precise role in the symbiotic relationship remained unclear. In 2017, Dr. Wang's team in China uncovered that these transporters transfer fatty acids—essential components of lipids—from host plants to symbiotic fungi [58]. The researchers suspected that the endomycorrhizal fungus *Rhizophagus irregularis*, widely used in agriculture and horticulture as a soil inoculant, depends on plants not only for sugars but also for lipids, because it lacks the genes required for fatty acid biosynthesis. To test this hypothesis, they conducted sophisticated isotope-labeling experiments, which confirmed that the fungus survives by obtaining fatty acids from its host plant. STR1 and STR2 are essential for symbiosis because they are the transporters that carry fatty acids from the host plant to symbiotic fungi.

Following this discovery, it was revealed that most microorganisms symbiotic with plants have lost their fatty acid synthesis genes, relying instead on plants as their lipid source. When symbiosis with fungi begins, plants increase fatty acid synthesis, likely to produce lipids for their fungal partners [105]. During this process, *WRINKLED1* (WRI1)-type plant transcription factors[9] enhance the expression of genes involved in lipid biosynthesis. Plants that fail to properly express the *WRINKLED1* gene cannot establish symbiosis with endomycorrhizal fungi. Similarly, liverworts, one of the earliest forms of terrestrial plants, show increased *WRINKLED1* expression when initiating symbiosis. Conversely, liverworts with artificially silenced *WRINKLED1* genes are unable to form symbiotic relationships with fungi.

After learning these stories, I began to wonder which nutrient plants provide to microorganisms is more important—sugars or lipids. I suspected lipids, since most endomycorrhizal fungi that symbiotically interact with plants lack fatty acid synthesis genes, meaning they must get lipids from another source. Moreover, lipids are energy-dense, providing roughly twice as many calories as sugars, making them a highly efficient energy source. Finally, fungi can convert fatty acids into sugars. At the International Workshop on Plant Membrane Biology held in Taiwan in late March 2023, I had the opportunity to discuss this question with Dr. Harrison from Cornell University, a leading researcher in this field. Dr. Harrison agreed that lipids are likely more critical for symbiosis than sugars, because she observed that mutants with defective lipid transporters cannot form symbiotic relationships with mycorrhizal fungi, while mutants with defective sugar transporters experience only a slight reduction in symbiosis. Why, then, have people long assumed that plants primarily supply sugars to symbiotic fungi, while the importance of lipids remained unknown? Perhaps it is simply because fewer scientists have studied lipids compared to sugars or proteins, leaving this area underexplored. After all, things can only be seen if one looks for them; otherwise, they remain unnoticed.

[9]A transcription factor is a protein that binds to specific DNA sequences to regulate gene expression by either increasing or decreasing the production of RNA from a gene.

Symbiosis Between Plants and Nitrogen-Fixing Bacteria

Next, let us explore the symbiosis between plants and nitrogen-fixing bacteria. Plants began forming symbiotic relationships with nitrogen-fixing bacteria around 100 million years ago. This symbiosis evolved independently multiple times, leading to a variety of mechanisms. Why did plants need to collaborate with nitrogen-fixing bacteria? This is because plants require proteins, and nitrogen is an essential building block of proteins. And, at the same time, readily available nitrogen-containing compounds are not abundant in most soils.

Plants use inorganic nitrogen sources, such as nitrates and ammonium, to produce proteins. To obtain these nitrogen compounds, plants establish cooperative relationships with fungi, rhizobia, *Frankia* bacteria, and cyanobacteria. Plants receive nitrates and ammonium from these microorganisms and, in return, provide sugars, organic acids, and a specialized environment for nitrogen fixation—an oxygen-free environment. Among these partnerships, the most familiar to people is the symbiosis between legumes and nitrogen-fixing bacteria. Many legumes modify parts of their structure to form root nodules, which create an ideal environment for nitrogen-fixing bacteria to reside in. Within these nodules, the plants cultivate rhizobia to secure their nitrogen nutrition.

Before delving into the highly specialized and complex structure of root nodules, let us first explore simpler and looser forms of symbiosis. Some plants host nitrogen-fixing bacteria within their internal spaces without forming specialized structures like root nodules. These plants include hornworts, liverworts, water ferns (such as *Azolla filiculoides*, a member of the Salviniaceae family), and certain gymnosperms like cycads (Fig. 4.6). These plants allow nitrogen-fixing

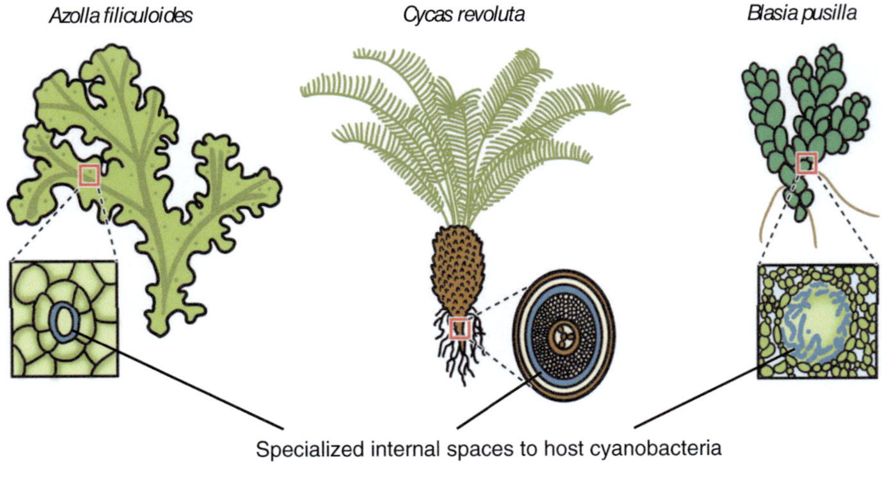

Fig. 4.6 Plants that host nitrogen-fixing cyanobacteria within their bodies. *Azolla filiculoides* (aquatic fern), *Cycas revoluta* (Sago palm—a gymnosperm), and *Blasia pusilla* (a liverwort) form specialized internal spaces to host cyanobacteria. These plants provide sugars to the cyanobacteria, stimulating them to fix more nitrogen

Fig. 4.7 Photograph of root nodules in soybeans

cyanobacteria to live in spaces within their leaves and absorb the nitrogen compounds fixed by these microorganisms. To support the cyanobacteria, the plants provide sugars necessary for their survival. At the same time, they secrete flavonoid[10] compounds that stimulate the cyanobacteria to fix more nitrogen. By effectively managing and guiding these microorganisms, the plants acquire the inorganic nitrogen (such as nitrate and ammonium) they need for growth.

Plants that form specialized structures called root nodules to house bacteria and obtain nitrogen compounds include various legumes and alder species. Legumes form symbiotic relationships with rhizobia (Fig. 4.7), while alders form symbioses with actinomycetes.

Formation of Root Nodules

How do plants and nitrogen-fixing bacteria establish their symbiotic relationship? Let us examine a representative example: the formation of root nodules in legumes and rhizobia. Below is a simplified explanation of this somewhat complex process.

As shown in Fig. 4.8, in step (A), legumes growing in nitrogen-deficient environments secrete flavonoid compounds from their roots, advertising their need for symbiotic bacteria. Rhizobia in the soil detect the flavonoids, realizing a legume root nearby is seeking a partner. The rhizobia then move toward the plant, attach to its root hairs, and produce signaling molecules called lipochitins, which are chitin[11] molecules with fatty acid groups. These chemical signals are unique to nitrogen-fixing bacteria, ensuring only these bacteria can form symbiotic relationships with plants for nitrogen fixation. The lipochitin signals bind to specific receptors on the

[10] Flavonoids are natural substances made by plants that belong to a group called polyphenols. One type of flavonoid, called anthocyanins, plays a key role in giving fruits their red, purple, or blue colors.

[11] A polysaccharide composed of sugar-derived amino compounds, serving as a major structural component of arthropod exoskeletons (shrimp and crabs), and fungal cell walls.

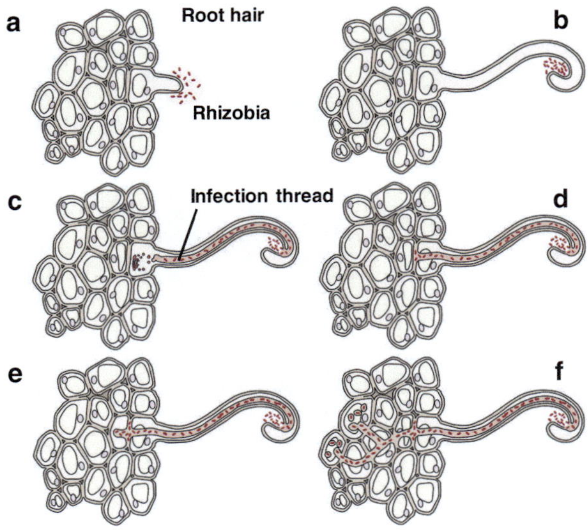

Fig. 4.8 Signal transduction process required to initiate symbiosis between legumes and nitrogen-fixing rhizobia. (**a**) Rhizobia recognize legumes secreting flavonoids and move toward the plant roots. The bacteria attach to the growing root hairs and produce lipochitins. (**b**) In response to lipochitins, the plant modifies the growth of its root hairs into a curled shape, within which the bacteria proliferate. (**c**) The cell wall of the root hair cell is modified to form an infection thread, allowing the bacteria to move inside. (**d**) The infection thread fuses with the cell membrane of the root hair cell, releasing the bacteria into the space between the cell walls. (**e**) Another infection thread is formed in cells located further inward. (**f**) The bacteria are released into the cytoplasm, where they live as part of the plant cell, functioning like cellular components

plant's cell membrane, activating them. Once activated, these receptors trigger changes in the plant necessary for symbiosis. As shown in step (B), the plant's root hairs curl, and cortex cells in the root dedifferentiate[12] and start dividing again. These changes reshape the root structure, allowing the bacteria to enter the plant cells. In this way, lipochitin acts as a key, unlocking the plant's cells for the bacteria.

In the next step (C), as the bacteria enter the curled root hair, a thin, thread-like structure called an infection thread forms. This structure arises from modifications to the plant's cell wall and cell membrane, creating a pathway for the bacterial entry. Inside the infection thread, as shown in (D) and (E), the bacteria grow and divide, while the infection thread extends into the root hair cell and into the cells below, delivering the bacteria into the plant cells. Subsequently, in step (F), the bacteria continue to grow and divide within the plant cells, expressing genes required for nitrogen fixation and differentiating into a symbiotic form optimized for efficient nitrogen fixation. Through this process, a specialized structure called the root nodule is formed, capable of housing large populations of nitrogen-fixing bacteria.

[12] Dedifferentiation is a process in which a plant cell that has specialized into a specific tissue loses its characteristics and reverts to an earlier cell state.

As described above, root nodules are formed through close communication between the plant and bacteria, enabling nitrogen-fixing bacteria to enter the interior of plant cells. While most microorganisms that establish intimate symbiosis with plants are fungi, and bacteria usually remain outside plant cells, nitrogen-fixing bacteria are an exception. These bacteria enter plant cells, proliferate extensively, and establish a unique symbiotic relationship with their host plant.

How Plants Manage Root Nodules

Legumes invest significant energy resources in creating the complex structure of root nodules and in hosting bacteria within the nodules. When nitrates, ammonium, or amino acids are available in the soil, absorbing these nutrients directly through the roots is far less resource-intensive than forming symbiotic relationships with rhizobia. Consequently, legumes in fertilized soils form few root nodules. Even if nodules have already formed in the absence of fertilizer, adding nitrogen fertilizer will cause the plant to stop supplying sugars to the nitrogen-fixing bacteria, leading to the death of the root nodules.

Rhizobia do not rely on symbiosis with plants for survival; they can live independently in the soil. When in symbiosis with plants, nitrogen-fixing strains thrive, while in nonsymbiotic soil environments, non-nitrogen-fixing strains tend to dominate and proliferate more successfully.

From the plant's perspective, root nodules function as factories that cultivate bacteria to fix nitrogen. In the initial stages of nodule formation, plants appear to accept both efficient and inefficient nitrogen-fixing bacteria without clear discrimination. However, in later stages, plants allocate fewer resources to bacteria that are not effective nitrogen fixers. In some cases, plants generate ROS in root cells heavily infiltrated by inefficient nitrogen-fixing bacteria, killing the bacteria. As a result, while ineffective nitrogen-fixing bacteria may enter root nodules, they do not proliferate actively. Nevertheless, plants do not entirely eliminate less efficient strains, and various strains of nitrogen-fixing bacteria, both effective and less effective, coexist within the plant. Why do plants not select only the most efficient nitrogen-fixing bacteria? Could it be that future environmental changes might make certain strains of bacteria more effective at nitrogen fixation? Or is it because evaluating the nitrogen-fixing efficiency of each bacterial strain in detail would require too many resources?

Exchange of Nutrients in Root Nodules

Similar to the exchange of substances observed in mycorrhizae, as illustrated in Fig. 4.5, nutrients are exchanged between plants and microorganisms in root nodules. In my laboratory, we previously studied transporter proteins critical to this exchange in root nodules [57]. To identify transporter genes involved in this process, we began by searching for genes expressed exclusively in alder root nodules and not in other plant tissues. Our rationale was that genes expressed solely in root nodules were likely to play a role in material exchange between the plant and rhizobia. One such gene was *AgDCAT1*. The amino acid sequence of the AgDCAT1 protein suggested that it spans the cell membrane multiple times, forming a structure characteristic of transporter proteins. Based on these observations, we hypothesized that AgDCAT1 is a transporter protein involved in material exchange within root nodules.

To test this idea and determine the function of this protein, we posed the question: Where within the structure of root nodules does *AgDCAT1* function, and what substances does it transport? To address this, we generated antibodies specific to the *AgDCAT1* protein and applied them to thin sections of root nodules, which we then examined under a microscope. The results showed that the AgDCAT1 protein localized to the symbiotic interface between alder root cells and the actinobacterium *Frankia*, enclosing the microsymbiont. To determine which substances *AgDCAT1* transports, we expressed the gene in yeast and *E. coli* and tested various candidate compounds. We found that *AgDCAT1* transported organic acids such as malate, fumarate, and succinate but did not transport nitrates or amino acids.

Based on these experimental results, our research team concluded that AgDCAT1 functions as a transporter, delivering malate from alder to *Frankia* bacteria. Malate, produced during photosynthesis, can be readily converted into sugars. We proposed that malate, synthesized during photosynthesis in alder, is transported across the biological membrane to *Frankia* bacteria via AgDCAT1, where the bacteria use it as an energy source for nitrogen fixation.

Using Nitrogen-Fixing Bacteria in Agriculture

If nitrogen-fixing bacteria can assist plants so efficiently, have there been attempts to apply them in agriculture? Experiments have been conducted to boost soybean yields by selecting the most efficient nitrogen-fixing bacterial strains and introducing them into soybean fields. However, reports indicate that these introduced bacteria were soon replaced by native bacterial species already present in the soil.

To address this issue, the Gates Foundation funded a research project called "N2Africa," aimed at optimizing soybean growth and field management by matching specific soybean varieties with the most compatible bacterial strains. Efforts have also been made to improve other crops to enhance their ability to form symbiotic relationships with nitrogen-fixing bacteria. In 2012, the Gates Foundation provided approximately ten million dollars to the John Innes Centre, a biotechnology research institute in the United Kingdom, to investigate how to enable corn to form symbiotic relationships with nitrogen-fixing bacteria. Corn is a vital crop for small-scale farmers in Africa. If corn could naturally obtain nitrogen through symbiosis rather than relying on synthetic nitrogen fertilizers, it would not only be more economical but also have significant environmental benefits.

To take it a step further, could crops be engineered to fix nitrogen directly without relying on symbiosis with bacteria? This is an extremely challenging goal. Nitrogenase, the enzyme that catalyzes the conversion of atmospheric nitrogen into ammonia, loses its activity even at slightly elevated oxygen levels. Given that plants produce large amounts of oxygen as a byproduct of photosynthesis, creating the low-oxygen environment within a plant's tissues is inherently difficult. Additionally, nitrogen fixation requires a substantial amount of energy.

How might future biologists tackle this challenge? A first step would likely involve engineering specialized structures capable of maintaining low oxygen levels. Since roots do not perform photosynthesis—which releases oxygen—they naturally maintain lower oxygen concentrations and may therefore be the most suitable site for such structures. Within these roots, scientists could construct nodule-like compartments and express genes that capture or sequester oxygen to further reduce oxygen levels. The next step would involve expressing genes for nitrogen-fixing enzymes, along with high-level expression of transporters to deliver photosynthetic products from the leaves as an energy source. Ultimately, this engineered structure would resemble natural root nodules—but without the bacteria. While this may seem like a daunting task, advancements in life sciences and biotechnology could one day make such an achievement possible.

5 The War Between Plants and Microorganisms

Let us now explore the interactions between plants and microorganisms that invade them. Because plants produce nutritious, high-calorie substances like sugars and fats, numerous microorganisms invade plants to access these resources. Crops, which accumulate more nutrients than plants living in natural environments, face even more intense microbial attacks. Despite significant human efforts to protect crops—such as using pesticides and developing disease-resistant varieties—an estimated 20–30% of global crop yields are still lost annually to pests and diseases [112]. While parallel statistics among wild plants is unknown, to survive such life-threatening attacks and reproduce, plants have evolved various strategies. Notably, when under attack from pathogens, plants typically halt growth and switch from a growth-oriented mode to a defense mode, reallocating resources accordingly.

Preemptive and Baseline Defense Strategies

Plants possess defense mechanisms that are prepared even before the invasion of pathogens. Reinforcing the cell wall to prevent pathogens from penetrating is a fundamental defense strategy. However, plants cannot create an entirely impenetrable barrier because they need to absorb and secrete various substances to survive. Another basic defense mechanism involves synthesizing and storing toxic compounds to protect themselves against pathogens. Plants produce an extensive array of defensive substances, including artemisinin in sweet wormwood, morphine in poppies, taxol in yews, nicotine in tobacco, tannins in persimmons, caffeine in coffee, and catechins in tea plants. Many of these compounds have been used by

Y. Lee, B. Y. Choi, *Social Lives of Plants*,
https://doi.org/10.1007/978-981-96-9837-0_5

humans for centuries as medicines. Artemisinin from sweet wormwood has been used as a treatment for malaria; morphine from poppies is a widely used painkiller, and taxol from yews has been employed in cancer therapy. Interestingly, many plant defense compounds are also popular stimulants or recreational substances enjoyed by humans. For instance, drinking coffee or tea can improve mood and increase alertness, potentially enhancing productivity. However, excessive consumption of these plant-derived stimulants can harm the body. An extreme case is tobacco smoking, which can be seen as humans falling victim to a plant's defense strategy, ultimately damaging their own health.

Do plants that produce abundant defensive compounds always have an advantage over others in nature? Not necessarily. Producing and storing defensive compounds is resource-intensive, and focusing on such efforts in the absence of pathogens can place these plants at a disadvantage compared to faster-growing competitors that invest more in growth and reproduction. For example, among various radish cultivars, the one most resistant to leaf spot disease caused by pathogenic fungi grows the slowest [51]. While plants that preemptively produce defensive compounds are more likely to survive when pathogens invade, such situations do not always occur. Furthermore, even plants that do not produce defensive compounds under normal conditions can often defend themselves by rapidly detecting pathogen invasion and activating defense mechanisms, such as synthesizing defensive compounds.

Plant Responses to Pathogen Invasion

When disease-causing microorganisms attack plants, their strategies can be broadly categorized into three types. First, necrotrophic (feeding on dead) pathogens invade plants, kill plant cells, and obtain nutrients from the dead tissue. Second, biotrophic (feeding on live) pathogens infiltrate plant cells without killing them, extracting nutrients over an extended period. Lastly, hemibiotrophic (half-biotrophic) pathogens initially behave like biotrophic pathogens, establishing themselves without killing plant cells, but later kill the plant to extract nutrients. Pathogens also differ in their host specificity. Some infect only specific plants, displaying host specificity, while others infect a wide range of plant species.

How do plants recognize and respond to such diverse pathogen attacks? Plants employ highly complex and varied methods to detect pathogen invasion and to protect themselves; they have multiple mechanisms to sense intrusion and, upon recognition, activate diverse defense responses. One common strategy is the production of ROS to kill pathogens, coupled with sacrificing infected plant cells to prevent the pathogen from spreading further. Another method involves synthesizing various chemical compounds for defense. Furthermore, upon infection, plants not only activate defense responses at the local site but also transmit signals to distal tissues to enhance resistance against potential subsequent infections, a process known as systemic acquired resistance (SAR). These strategies help plants eliminate invading pathogens and better resist future attacks.

How Plants Detect Pathogen Invasion

How do plants quickly detect the invasion of pathogens? Extensive research by numerous plant biologists has provided insights into this process. Plants recognize pathogen intrusion by detecting molecules known as microbe-associated molecular patterns (MAMPs), which originate from pathogens, or damage-associated molecular patterns (DAMPs), which are released by damaged or dying plant cells due to pathogen attack (Fig. 5.1).

A prime example of a MAMP is flagellin, a peptide derived from the flagella[1] of microorganisms. Many pathogenic microorganisms possess flagella, which enable them to move efficiently and infect plants. Consequently, fragments of flagella can serve as key signals to plants, indicating the presence of motile pathogens. Like flagellin, many MAMPs are specific to microorganisms and play critical roles in their survival or infection processes.

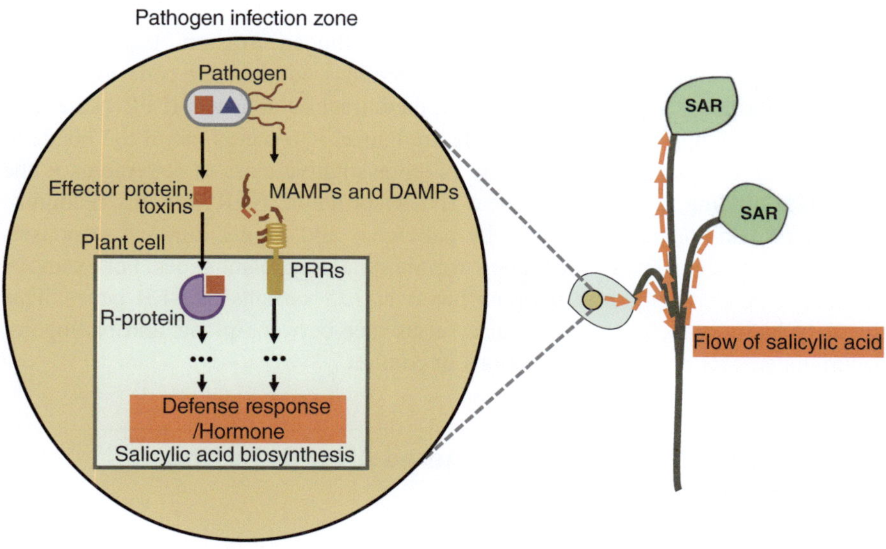

Fig. 5.1 Plant immune responses to pathogen invasion. The pattern recognition receptor (PRR) proteins on the plant cell membrane detect microbe-associated molecular patterns (MAMPs) originating from pathogens and damage-associated molecular patterns (DAMPs) produced during the invasion process and thus identify the presence of pathogens outside the cell. Inside the cell, R proteins recognize effector proteins and toxins released by the invading pathogens. When activated by the detection of pathogens, PRRs and R proteins initiate signal transduction pathways that trigger various defense responses. These include the production of defense hormones such as salicylic acid. Salicylic acid, synthesized at the site of pathogen infection, moves to uninfected plant tissues (depicted by the orange arrows), promoting systemic acquired resistance (SAR). This enhances resistance to pathogens throughout the plant

[1] Flagella are long, whip-like appendages protruding from a cell, providing motility to the cell.

Another example of MAMPs is chitooligosaccharides, which are released when the cell walls of fungi or bacteria are disrupted. Their presence indicates pathogen cell wall damage and suggests that the pathogen has invaded the host. However, chitooligosaccharides are also known to play an essential role in forming symbiotic relationships. If chitooligosaccharides can signify both the presence of a friend and a foe, how do plants decide how to respond? At the end of Chap. 6, we will explore how plants differentiate between the chitooligosaccharides released by symbiotic fungi and pathogenic fungi.

When MAMPs and DAMPs are present, how do plants recognize them? The proteins responsible for monitoring the presence of MAMPs or DAMPs on the cell membrane are called pattern recognition receptors (PRRs). When MAMPs or DAMPs bind to these receptors, the receptor undergoes a structural change, sending a signal into the cell to indicate a pathogen invasion. In response, the cell activates various defense mechanisms. Genes encoding PRRs that detect chitooligosaccharides are found not only in vascular plants but also in mosses, which can recognize and respond to chitin. This indicates that pathogen detection through PRRs likely evolved early in plant evolution.

In addition to PRRs that detect evidence of pathogen invasion outside the cell, plants also possess resistance proteins (R-proteins) within their cells to monitor whether microbial molecules have entered the intracellular space. If PRRs are akin to guards watching for intruders outside the house, R-proteins patrol the inside to check if a thief has already broken in. A representative class of R-proteins is the nucleotide-binding leucine-rich repeat (NLR) proteins. NLR proteins recognize effector proteins or toxins secreted by pathogens and trigger immune responses. NLR genes exhibit some of the highest variability among plant genes. For example, the model plant *Arabidopsis thaliana* has over 125 variants of NLR genes. This diversity likely reflects the evolutionary arms race between plants and pathogens, accumulated over countless generations of conflict.

Plant Defense Mechanisms Triggered After Detecting Pathogen Invasion

PRRs and R-proteins activated by pathogen invasion initiate a variety of defense responses as outlined below.

First, infected plant cells and their neighboring cells produce large amounts of ROS, triggering a hypersensitive response (HR) that rapidly kills infected cells and the pathogens within them (Fig. 5.2). This response prevents the pathogen from spreading to uninfected cells. In the surrounding cells, lignin[2] synthesis increases, and cell wall proteins are modified, resulting in a strengthened cell wall. These reinforced walls trap any remaining pathogens within the dead cells, preventing their spread to other parts of the plant.

[2] A complex phenolic polymer that reinforces the plant cell walls, providing structural support in woody plants.

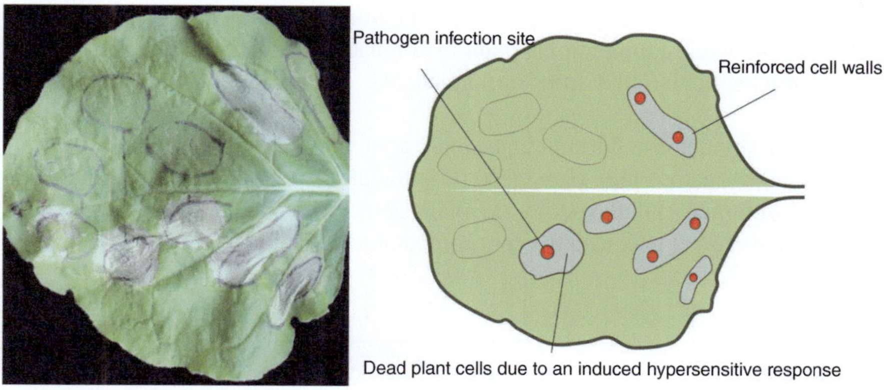

Fig. 5.2 Hypersensitive response triggered in plant cells by pathogen infection. A photograph (left) and its corresponding illustration (right) demonstrate the hypersensitive response (HR) in plant cells. Tobacco leaves were inoculated with *Agrobacterium* engineered to express various pathogen-derived effector proteins. The inoculated areas are outlined with black lines. The gray areas of the leaf were where the HR occurred, which killed tobacco leaf cells. In contrast, areas that retained their original color are regions where HR was not triggered. Such experiments are instrumental in identifying which pathogen-derived effector proteins induce HR in tobacco leaves. The plant cells surrounding those killed by the HR develop reinforced cell walls, effectively restricting the spread of the pathogen

Second, plants increase the synthesis of phytoalexins to attack and eliminate invading pathogens. Phytoalexins belong to the group of toxic chemical defense compounds previously described. Many plants produce different types of phytoalexins in response to attacks, and use these compounds to protect themselves.

Third, plants enhance the synthesis of pathogenesis-related (PR) proteins to combat invading pathogens. PR proteins encompass a broad range of proteins produced in response to pathogen invasion. Among these are hydrolytic[3] enzymes, such as chitinases, which break down chitin, a key component of fungal cell walls. By degrading fungal cell walls, these proteins inhibit fungal growth and improve the plant's disease resistance.

Finally, plants respond by increasing the synthesis of their defense hormones, salicylic acid and jasmonic acid. The immune responses mediated by these defense hormones vary greatly depending on the characteristics of the pathogen, making them difficult to explain in simple terms. However, without these defense hormones, the plant's defensive responses become significantly weakened, underscoring the importance of understanding their general roles.

The immune responses activated by salicylic acid are particularly crucial in protecting plants from attacks by biotrophic pathogens. Salicylic acid, an essential oil with a wintergreen scent and chemically similar to the fever-reducing aspirin, is synthesized at the site of pathogen infection and spreads to uninfected tissues

[3] Hydrolysis is a chemical reaction in which a large molecular compound reacts with water molecules and breaks down into smaller ions or molecules.

(Fig. 5.1). This triggers defense responses and induces systemic acquired resistance (SAR) in 3–5 days, equipping the plant to combat subsequent pathogen infections. In this way, salicylic acid functions much like a vaccine, fortifying the plant's defenses.

Another key defense hormone, jasmonic acid, plays a crucial role in protecting plants against necrotrophic pathogens. It stimulates the biosynthesis of anthocyanins and antimicrobial flavonoids in plant cells, which help suppress pathogen growth. Herbivore attack also triggers comparable jasmonic acid-mediated defense responses.

Strategies of Pathogens Invading Plants

Just as plants have developed various methods to defend themselves, pathogens have evolved strategies to successfully invade plants. These strategies involve either disabling the plant's immune responses or creating a favorable environment within the plant for the pathogen to thrive and reproduce. One method pathogens use to neutralize plant immune responses is to inject effector proteins into plant cells (Fig. 5.1). Pathogenic bacteria typically inject approximately 15–30 effector proteins into plants. To understand whether these effector proteins are truly essential for infection and what roles they play inside the plant cells, scientists have conducted extensive research.

In one experiment, effector protein genes in pathogens were mutated[4] to stop the delivery of effector proteins into plant cells. Compared to wild-type pathogens, these mutants showed a significantly reduced ability to infect plants, demonstrating that effector proteins play a critical role in facilitating infection. Ongoing research aims to uncover the specific mechanisms by which effector proteins benefit pathogens.

One such mechanism involves the pathogen's ability to proliferate quickly during the early stages of infection. To establish a successful infection, the pathogen must multiply rapidly and reach a high population level before the plant detects its presence and activates defenses. If the plant detects the pathogen early, it can greatly restrict its replication. To support rapid growth, the pathogen draws nutrients from plant cells—a process that relies heavily on effector proteins. For example, the TAL effector protein increases the expression of the plant's sugar transporter genes, known as SWEET [13]. Pathogens unable to enhance the expression of these sugar transporter genes are less effective at infecting plants. Under normal conditions, SWEET proteins function as sugar transporters in the plant cell membrane, facilitating the secretion of sugars outside the cell. They play essential roles in processes like nectar secretion in flowers and sugar storage in fruits and are typically expressed only in specialized tissues where sugar is needed. However, when pathogens invade the plant, they can induce *SWEET* expression at infection sites where it is not normally expressed. This causes sugars to be secreted from plant cells via SWEET

[4] A mutation is a change in the DNA sequence that constitutes genetic information.

transporters, making these sugars accessible to the pathogens. Essentially, the pathogens exploit the plant to obtain the resources they need to quickly multiply. This scenario is akin to receiving supplies from the enemy during a war, making it far easier for the invader to achieve victory.

Some pathogens secrete plant hormones to facilitate invasion. In Japan, rice farmers have long observed a phenomenon where rice plants grow excessively tall, topple over, and fail to produce seeds, calling them "foolish seedlings." Studies on this phenomenon in the late nineteenth century by Japanese researchers discovered that the pathogenic fungus *Gibberella fujikuroi* secretes a substance that causes excessive elongation in rice plants. They named this substance "gibberellin" after the fungus. Subsequent research uncovered the astonishing fact that gibberellin is actually a plant growth hormone naturally synthesized in healthy plants— albeit in extremely small amounts—to promote normal growth [48]. The ability of a fungus to produce a plant hormone to attack its host is truly remarkable. Unlike the controlled production of gibberellin in plants, the *Gibberella* fungus secretes large quantities of the hormone, causing the plant to grow abnormally tall and weak, with thinner (and more easily penetrable) cell walls, ultimately leading to its death. Similarly, many pathogens produce another plant hormone, cytokinin, which has antiaging effects. If the infected plant tissue were to die and dry out, the pathogen would lose its source of nutrients. By secreting cytokinin, pathogens delay plant senescence, enabling them to extract nutrients from the plant for an extended period.

The Moisture War Between Plants and Pathogens

Now let us examine how pathogens create a favorable environment within the plant to support their survival and reproduction and the role of effector proteins in this process. Some effector proteins increase moisture levels inside plant leaves, thereby creating conditions that promote pathogen growth and propagation [138]. This finding emerged from studies exploring why crops are more prone to disease during periods of heavy rainfall or high humidity. The research focused on *Pseudomonas syringae*, a pathogenic gram-negative bacterium that infects tomatoes and *Arabidopsis thaliana*. They found that at a high humidity level of approximately 95%, the pathogen had a 95% chance of infecting its host plant. However, when the humidity was comparatively lower, at 40–69%, the infection rate dropped to 60%. One of the first symptoms observed in plants infected by this pathogen was water-soaked areas within the leaves, which later corresponded almost exactly to the regions where the disease developed. This led researchers to question whether effector proteins were responsible for inducing this water-soaked appearance in plant tissues. To test this hypothesis, they created a mutant strain of *Pseudomonas syringae* (Pst-DC3000-28E) that could not express any of its 28 effector protein genes. When this mutant was introduced to plants, even under high humidity, it failed to produce the water-soaked appearance in plant tissues, confirming that effector proteins play a key role in this process.

The next step was to determine which of the 28 effector proteins caused this effect. To identify the specific proteins, the researchers reintroduced the effector protein genes one by one into the mutant strain of the pathogen, which was unable to produce any effector proteins, and then inoculated plants with the modified pathogen. The results showed that when *Pseudomonas syringae* expressing the *HopM1* and *AvrE1* genes was introduced to plants, the water-soaked appearance in the leaves returned. Conversely, deleting both *HopM1* and *AvrE1* genes from the wild-type *Pseudomonas syringae* strain (Pst-DC3000) prevented the pathogen from inducing water-soaked symptoms, even under high humidity, and rendered it unable to cause disease.

Further research has shown that AvrE-family effectors function as water- and solute-permeable channels [90]. These channels disrupt osmotic and water potential conditions in the infected plant tissues, creating a water- and nutrient-rich space between plant cells where bacteria can proliferate.

How does moisture make it easier for pathogens to infect plants? First, abundant moisture in its liquid form dilutes the defense compounds secreted by plants, lowering their concentration. For example, the levels of plant defense hormones like salicylic acid and jasmonic acid cannot rise sufficiently, and their signaling pathways fail to activate effectively, weakening plant immunity. Second, excess moisture facilitates the uptake of nutrients by pathogens from the plant. When nutrients dissolve in water, it becomes easier for transporter proteins on the pathogen's cell membrane to carry these nutrients into their cells. Third, increasing water content within the plant disrupts the homeostasis[5] maintained by the plant's friendly endophytic microorganisms, creating conditions more susceptible to disease [138].

The story does not end here. The relationship between plants and their pathogens is constantly evolving. For every strategy employed by pathogens, plants develop counterstrategies, which bacteria then try to overcome. Thus, it is no surprise that there are many stories about moisture in plant-pathogen interactions. Here are more of those stories. The stomata on plant leaves are not only pathways for water and carbon dioxide but also potential entry points for pathogens. Guard cells, which regulate the size of the stomatal openings, recognize flagellin fragments like flg22 and respond by closing the stomata (Fig. 5.3). When the stomata close, pathogens find it difficult to penetrate the leaf through these openings. But what happens if the pathogens have already entered the leaf? Once inside, bacteria manipulate the stomata to remain closed, preventing water from escaping and thereby increasing the humidity inside the leaf—a favorable environment for their proliferation. This process involves hijacking the plant's abscisic acid (ABA)[6] signaling pathway, which

[5] Homeostasis is the process by which an organism or cell maintains a stable internal state despite external and internal changes.

[6] ABA is a plant hormone that builds up when a plant is under stress such as drought. It helps protect the plant by closing tiny pores on the leaves (called stomata), slowing growth, and triggering various other stress responses.

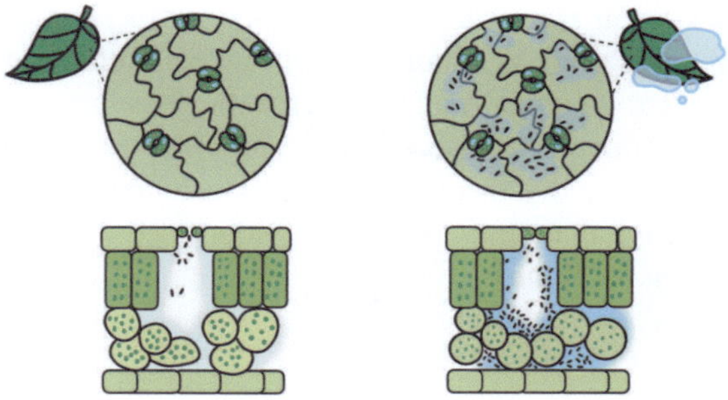

Stomata	Open	Closed
Transpiration rate	High	Low
Humidity within leaves	Low	High
Concentration of defense molecules	High (concentrated)	Low (diluted)
Homeostasis of endophyte	Maintained	Disrupted
Plant immunity	High	Low

Fig. 5.3 Pathogens increase leaf humidity, facilitating proliferation. When ambient humidity is high or stomata are closed, the internal humidity of leaves rises, creating ideal conditions for pathogen growth. This helps explain, at least in part, why plant diseases are more common during humid summers. *Pseudomonas* bacteria promote these conditions by inducing stomatal closure, reducing transpiration, and preventing water loss from the leaf—thus fostering an environment conducive to their proliferation

the plant typically uses to close stomata in response to drought [52, 109]. The pathogens exploit the plant's signaling processes in a way that resembles a thief using the homeowner's defense equipment for their own purposes.

To counteract pathogens that increase internal humidity, plants adopt an opposite strategy: they keep their stomata open to reduce leaf humidity [77]. A research team studying this phenomenon identified four plant peptides induced by flg22 and named them SCREW peptides. These peptides inhibit the ABA signaling pathway, keeping the stomata open. This allows transpiration to continue, enabling water to escape from the leaf, reducing humidity, and creating conditions less favorable for pathogen infection.

Plants and their pathogens are continuously evolving, each trying to overcome the other's defenses. For both, survival is at stake, making it impossible to stop this evolutionary arms race. If any one plant or any one pathogen grows weary and gives up, it faces the risk of extinction. In natural ecosystems, plants and their pathogens seem to maintain a delicate equilibrium, where neither fully overcomes the other. In agriculture, however, pathogens that threaten crops are largely kept in check through the use of pesticides. Meanwhile, with the Earth's environment changing rapidly, there is growing concern that this equilibrium may be disrupted in nature and that pesticides may no longer be sufficient to control pathogens effectively in the future.

How Do Plants Distinguish Between Beneficial and Harmful Microorganisms?

6

Which Comes First: Symbiosis or Defense?

It is fascinating how plants identify the microorganisms they encounter, choosing to collaborate with some while launching vigorous defenses to repel others. The key to combating invasive pathogens lies in swiftly recognizing them as threats. Once the pathogen population becomes too large, activating defense mechanisms is often futile. Therefore, plants must determine whether a microorganism is a pathogen early on, when information is scarce. Under laboratory conditions, it has been shown that plants recognize pathogens within minutes to hours. During evolution, did plants prioritize symbiosis to secure nutrients, or did they first develop defense mechanisms to protect themselves from pathogens? While this life-or-death question remains largely unanswered, the text below summarizes the insights uncovered thus far.

Based on a comparative analysis of symbiosis-related plant genes across various photosynthetic organisms, Dr. Delaux's team in France proposed that the genes essential for symbiosis with microorganisms were already present when plants lived in aquatic environments. This suggests that plants were prepared for microbial symbiosis even before expanding their habitat to land [26, 27]. Furthermore, the scientists argue that the establishment of mechanisms for microbial symbiosis likely preceded the development of defenses against pathogenic invasions. While it is impossible to precisely determine the events of evolution from hundreds of millions of years ago, this explanation appears plausible. After all, pathogenic microorganisms, like other organisms, rely on plants to produce organic matter. If terrestrial plants could not acquire nutrients from the soil without symbiotic microorganisms and, as a result, could not survive, plant pathogens would have no hosts to infect. This argument aligns well with the evidence.

Arabidopsis thaliana suppresses its defense mechanisms under phosphate-deficient stress conditions [18]. This observation suggests that securing the minimum nutrients necessary for survival takes precedence over defense for plants.

Y. Lee, B. Y. Choi, *Social Lives of Plants*,
https://doi.org/10.1007/978-981-96-9837-0_6

Thus, it can be inferred that symbiosis with microorganisms to acquire nutrients is more fundamental and critical than defense against pathogens. It is akin to how soldiers in a war need a steady supply of food to sustain their strength and continue fighting. In conclusion, it seems plausible that plants first established symbiosis with microorganisms and later developed mechanisms to recognize and eliminate pathogens when they invaded.

If pathogens invaded plants by exploiting structures that originally evolved for symbiosis, then the initial strategies used by both pathogenic and symbiotic microorganisms to approach early plants may not have differed much. In that case, early plants forming symbiotic relationships with microorganisms likely faced challenges in defending themselves against pathogenic invasions. Over the course of 400 million years of evolution, however, plants had to adapt, developing the ability to recognize pathogens and mount defenses while continuing to cooperate with beneficial symbiotic microorganisms. Only plants capable of solving these two difficult problems have survived.

What changes may have occurred? The mechanisms to distinguish between beneficial and harmful microorganisms became more sophisticated, and the defense responses to eliminate pathogens became more selective and potent. In response, pathogens evolved strategies to evade or overpower plant defenses, while symbiotic microorganisms devised more distinct approaches to avoid being mistaken for pathogens. Through this prolonged struggle for survival, the weapons and strategies developed by both sides inevitably became increasingly complex and specialized [143].

How Symbiotic Microorganisms Evade Plant Defense Mechanisms

Symbiotic microorganisms share many traits with pathogens, raising the question: what unique strategies have they evolved to avoid being mistaken by plants for harmful invaders? Recent studies have shed light on how these symbionts suppress or evade plant defense responses [127]. While pathogens invading plants often use the type 3 secretion system (T3SS) to inject toxic substances into the plant, many symbiotic microorganisms establish relationships with plants by employing the type 2 secretion system (T2SS) (Fig. 6.1). T2SS system introduces immunosuppressive compounds into the plant roots, suppressing the plant's defense responses. An intriguing phenomenon has been observed: the presence of bacteria capable of suppressing or evading plant defenses facilitates the attachment of other bacteria, which otherwise could not establish symbiosis. In other words, bacteria that breach plant defensive barriers effectively create opportunities for others to coexist with the plant.

Another common strategy among symbiotic microorganisms is the loss of enzymes required to degrade plant cell walls—a clear contrast to pathogens, which often invade by breaking down these structures. By avoiding the release of cell wall fragments, which can trigger plant immune responses, symbionts minimize the risk of activating plant defenses.

To further understand how symbiotic microorganisms evade plant defense, Dr. Dangl's team in the United States studied flagellin22, a fragment of the bacterial

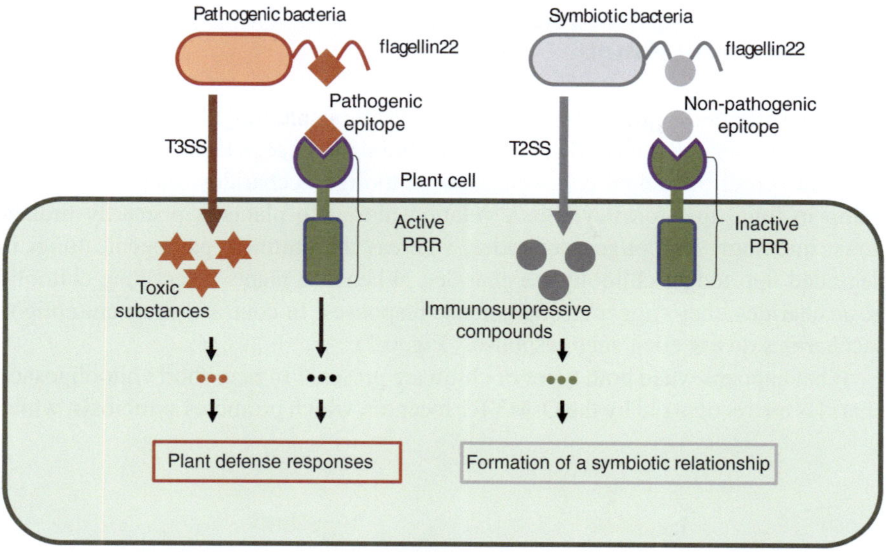

Fig. 6.1 Strategies used by symbiotic bacteria to evade plant immune responses, in contrast to pathogenic bacteria. Pathogenic bacteria utilize the T3SS to inject toxic substances into plant cells, while symbiotic bacteria use the T2SS to deliver immunosuppressive compounds. Toxic substances introduced via the T3SS are detected by the plant immune system, triggering defense responses. In contrast, immunosuppressive compounds delivered through the T2SS do not provoke such responses. The flagella of pathogenic bacteria contain epitopes that bind to and activate PRRs in the plant cell membrane, inducing defense responses. Conversely, the flagella of symbiotic bacteria carry epitopes that do not bind to PRRs

flagellum. Flagella enhance microbial motility, making it easier for microorganisms to penetrate plants; hence, both pathogens and symbiotic microorganisms often possess flagella. Among the structural components of flagella, flagellin22 is a critical segment that, when introduced to plants, triggers strong defense responses. The researchers explored how symbiotic microorganisms, which carry such a potent defense elicitor, manage to evade plant defenses and establish symbiotic relationships.

Their findings revealed that not all flagellin22 structures are identical; they vary among microorganisms. The flagellin22 of symbiotic microorganisms has different epitopes—the antigen-binding regions—compared to those of pathogens (Fig. 6.1). Because of this difference, the flagellin22 of symbiotic microorganisms does not activate the plant's PRRs. When PRRs are not activated, one of the critical components in the plant's immune responses to detect pathogen invasion fails to occur [22]. Consequently, the likelihood of the plant initiating defense responses decreases.

Interestingly, under conditions where soil salinity increases and plants are stressed or unhealthy, microorganisms with flagellin22 epitopes that activate PRRs tend to proliferate around the plant roots. This suggests that healthy plants actively create a root environment favorable to symbiotic microorganisms. In contrast, stressed plants fail to attract symbiotic microorganisms effectively, allowing pathogenic microorganisms to gather around the roots.

How Rice Distinguishes Between Symbiotic and Pathogenic Fungi

Rice can differentiate between symbiotic and pathogenic fungi by recognizing differences in their cell walls. Many plants, including rice, secrete enzymes that break down the chitin in fungal cell walls into chitooligosaccharides. Interestingly, the chitin in fungi that form symbiotic relationships with plants is primarily broken down into short chitooligosaccharides, whereas the chitin in pathogenic fungi is degraded into longer chitooligosaccharides. When rice plants detect long chitooligosaccharides, they trigger strong defense responses. In contrast, short chitooligosaccharides do not elicit such responses (Fig. 6.2).

What happens when both types of chitin are present? In rice, short chitooligosaccharides are recognized by the OsMYR1 receptor, which promotes symbiosis, while

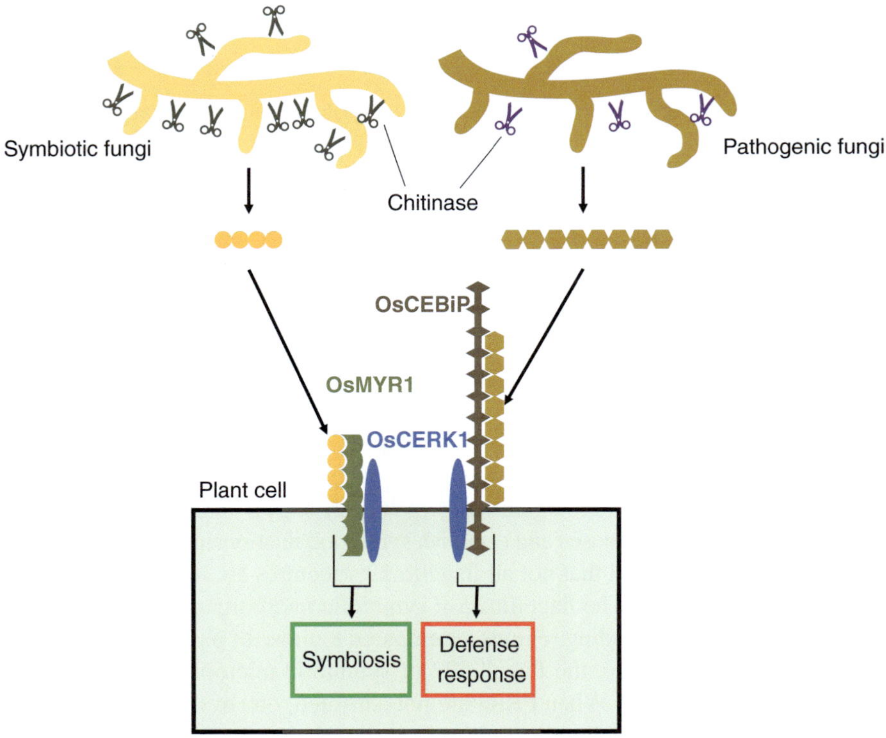

Fig. 6.2 A method used by rice to distinguish between symbiotic and pathogenic fungi. Rice breaks down fungal cell walls into chitooligosaccharides. The chitin cell walls of fungi that form symbiotic relationships with rice are primarily broken down into short chitooligosaccharides, whereas the chitin of pathogenic fungi is degraded into long chitooligosaccharides. Short chitooligosaccharides are recognized by OsMYR1, a receptor that promotes symbiosis. In contrast, long chitooligosaccharides are detected by OsCEBiP, a receptor that triggers defense responses. Both bound receptors compete for binding to the OsCERK1 receptor, the result—symbiosis or defense response—depends on the type that prevails

long chitooligosaccharides are recognized by the OsCEBiP receptor, which triggers defense responses (Fig. 6.2). Both receptors must bind to a signaling protein called OsCERK1 to activate their respective plant responses. If one receptor binds to OsCERK1 in large quantities, it reduces the number of binding sites available for the other receptor. Signals from receptors that fail to bind to OsCERK1 are not transmitted, and the responses they would induce do not occur. This competition for OsCERK1 binding takes place downstream of chitin recognition. In other words, in rice, receptors for two entirely different signals—accepting symbiotic fungi or repelling pathogenic fungi—compete for access to a single protein. Only the winner transmits its signal to the next stage. This signaling system ensures that when one response is triggered, the other is suppressed, allowing rice to decide whether to foster a symbiotic relationship or activate defenses against harmful fungi [146].

The validity of this explanation was further confirmed through the following experiments. Rice plants genetically engineered to overexpress the symbiotic receptor OsMYR1 showed improved symbiosis but were suppressed in defense responses and became more susceptible to diseases when exposed to pathogenic fungi. Conversely, suppressing the expression of the OsMYR1 gene enhanced disease resistance. Similarly, rice plants with artificially reduced expression of the defense receptor OsCEBiP formed symbiotic relationships with fungi more effectively.

A Method Based on the Degree of Physical Damage During Interactions

Unlike pathogens, symbiotic microorganisms rarely cause physical damage to plant cells. Dr. Geldner's team in Switzerland reported that plants distinguish between pathogens and symbiotic microorganisms based on the degree of physical damage they experience during interactions with the microbes [147]. They hypothesized that plants become sensitive to microorganisms and trigger defense responses only when sustained physical wounds occur, and it is why plants do not constantly activate their defense responses but only when pathogens are present.

To test this hypothesis, the researchers conducted delicate experiments: They used a laser to wound a few cells in the outer epidermal layer of plant roots. This procedure triggered a sharp increase in the expression of PRRs in neighboring cells, making the plant more responsive to defense induction. Based on this and other results, they explained the following: Under normal conditions, plant cells do not react to nearby microorganisms, even though such microbes are constantly present. However, when adjacent cells are wounded, neighboring cells rapidly increase PRR production, enhancing their ability to detect microbes. If microbial fragments—such as flagellar proteins or cell wall components—are then detected by these PRRs, a vigorous defense response is launched. Because pathogens often damage root cells in the early stages of infection, plants only enter this heightened state of alert upon receiving wound signals. This strategy appears effective, as remaining in a constant state of vigilance would consume valuable resources, hinder plant growth, and deter beneficial symbiotic microorganisms from residing near the roots.

Another remarkable finding was that when plant cells are wounded, they also trigger defense responses against microorganisms that were previously in symbiosis with them. Once damaged, the plant appears to perceive all microorganisms with similar characteristics as threats, attempting to block their entry and eliminate them. In short, wounding acts as the first gateway, and the detection of substances released by microorganisms further activates the defense responses in plants. Thus, plants seem to distinguish between pathogens and symbiotic microorganisms in their environment based on the degree of physical damage they experience during interactions with these microbes.

In addition to the mechanisms already discussed, there are likely many other ways plants distinguish between symbiotic and pathogenic microorganisms. A plausible example would be symbiotic microorganisms secreting unique signaling molecules that the host plant can recognize. However, pathogens could evolve over time to mimic these unique molecules, forcing symbiotic microorganisms and plants to continuously evolve and replace these signals with new ones—similar to how we regularly update passwords for online accounts. Since both acquiring nutrients through symbiosis and defending against pathogens are critical for plant survival, it is certain that plants have meticulously refined their mechanisms to excel at both over the course of long evolutionary processes. This suggests that the mechanisms underlying symbiosis and defense are likely to be incredibly diverse, and many new discoveries in this area are bound to emerge in the future.

In Part II, we explored the social interactions plants maintain with a variety of microorganisms. In summary, plants live in close association with microorganisms, collaborating extensively with beneficial ones. However, some microorganisms attack plants, forcing them to engage in defensive battles. Plants that have succeeded in both aspects of their social interactions—cooperating with beneficial microorganisms and defending against harmful ones—are the ones that have survived in nature. Evidence of these long-standing dynamics, which date back to the emergence of terrestrial plants, can be seen in the numerous symbiosis-related and defense-related genes present in plants today.

Scientists studying the relationship between plants and microorganisms aim to develop new crops that can effectively form symbioses with beneficial microbes while repelling harmful ones. To improve symbiosis, it is necessary to identify the key genes involved in the symbiotic process in both crops and microorganisms. In the past, researchers approached this problem by isolating individual microorganisms and analyzing their genes. However, with advancements in metagenomics, it is now possible to simultaneously sequence the genes of numerous microorganisms found in soil or living on plants. By analyzing the resulting big data in conjunction with plant growth and crop yield, scientists can shift the question from "Which microorganisms benefit crops?" to "Which genes in microorganisms benefit crops?" Taking this further, researchers can identify which host plant genes interact with specific genes in beneficial microorganisms. Advances in automated methods for measuring plant traits now allow researchers to analyze changes in plant growth under specific microbial conditions without extensive manual effort. Coupled with the increasing ease of plant genome analysis, this progress enables large-scale experiments that were previously impossible, offering new insights into the complex interactions between plants and microorganisms.

Understanding how plants form symbiotic relationships with microorganisms and applying this knowledge to crops could facilitate symbiosis, thereby improving their nutrient acquisition. This would reduce the need for excessive fertilizer use, promoting sustainable agriculture and mitigating the environmental pollution caused by fertilizers. Furthermore, advancements in synthetic biology techniques offer the possibility of modifying microbial genes or engineering new microorganisms to facilitate symbiosis with plants. Similarly, uncovering how plants defend themselves against pathogenic microorganisms and leveraging this knowledge could lead to the development of disease-resistant crops that require less pesticide application. Therefore, research into the social interactions between plants and microorganisms has the potential to contribute to the future of agricultural development.

Part III

Plants and Animals

In nature, plants face attacks from many different animals. However, this does not mean that plants are merely prey to animals. On the contrary, some plants use animals for reproduction or defense and, in special cases, even capture and digest them to obtain nutrients. Similar to the interactions between plants and microorganisms, the relationship between plants and animals has evolved over a long period, characterized by fluctuating dynamics. Occasionally, these dynamics shift dramatically, leading to changes in their interplay.

Animals that feed on plants are diverse, with insects and large vertebrate herbivores being the most studied examples. While both insects and large herbivores consume plants, their differences in size and behavior necessitate distinct approaches and focuses in research. Let us explore and compare the various social interactions plants form when encountering insects and large vertebrate herbivores.

The Relationship Between Plants and Insects

7

Insects That Aid Plant Reproduction

Approximately 75–80% of flowering plants[1] rely on animals to produce offspring. For flowering plants, pollen must be transferred from one flower to another to generate seeds, which begin the next generation. In most cases, this transfer is carried out by animals. About 200,000 animal species play the role of pollinators, the majority of which are insects [88]. Bees, such as honeybees, bumblebees, and wasps, as well as butterflies, moths, flies, ants, and beetles, are among these pollinators (Fig. 7.1). Butterflies are generally active during the day, while moths are active at night. Flowers that rely on moths often emit strong fragrances and are frequently white, making them easier for moths to locate at night. When insects recognize signals such as a flower's color or fragrance and visit to feed on nectar, pollen sticks to their bodies. As the insect moves from one flower to another, the pollen from the first flower can adhere to the stigma of the second, facilitating pollination. In pollinated flowers, fertilization happens when sperms developed inside the pollen unite with egg cells in the pistil, the female part of the flower. This leads to the creation of offspring.

When did plants begin to recruit insects for reproduction? Both flowering plants and insects diversified into numerous species during the mid-Cretaceous period, around 100 million years ago, with insect-mediated pollination playing a key role in these diversifications. Fossil studies provide evidence that insects facilitated plant pollination as far back as 99 million years ago, as seen in amber[2] fossils [10]. One

[1] Flowering plants (angiosperms) are seed-producing plants that bear flowers and fruits. Unlike gymnosperms—woody, perennial plants like conifers and cycads that produce exposed seeds—angiosperm seeds are enclosed within fruits.

[2] Fossilized tree resin, usually golden or brown, sometimes containing insects. Trees secrete resin to protect wounds and trap invading insects, which can become preserved as the resin hardens over millions of years.

Y. Lee, B. Y. Choi, *Social Lives of Plants*,
https://doi.org/10.1007/978-981-96-9837-0_7

Fig. 7.1 Insects that transport pollen. Clockwise from the top left: ground bee, honeybee, moth, and butterfly

such fossil preserved a previously unknown beetle from the family Mordellidae along with numerous pollen grains from various flowering plants, all trapped in amber. The authors of the study proposed that these pollen grains adhered to the large hind legs of the beetle and were transported to other flowers, suggesting that plants and insects had formed a mutually beneficial relationship as early as 99 million years ago. Thus it seems highly likely that, over the long course of history, from the past to the present, flowering plants have flourished through their interactions with insects.

Butterflies and moths play a crucial role in transferring pollen, but their larvae often act as pests that devour plants. While adult insects forage among flowers, their larvae, unable to travel between multiple flowers, feed on their host plants. For instance, the larvae of the cabbage white butterfly are pests to plants in the Brassicaceae family, causing considerable damage [95].

Some insects depend exclusively on specific plants. The larvae of the monarch butterfly feed only on the leaves of milkweed plants, such as *Metaplexis japonica*. Consequently, the population of monarch butterflies is directly proportional to the abundance of milkweed plants. When agricultural methods became available to remove weeds including milkweed plants in corn and soybean fields in the central United States, the population of monarch butterflies also declined [47].

While some insects rely exclusively on specific plants, there are more insects that gather nectar and pollen from various plants. Honeybees, for instance, visit multiple plants to collect nectar and pollen. They are essential pollinators for over 100 species of fruit and vegetable crops. Recently, the number of honeybees has been declining, though the exact cause remains unclear. Factors such as parasites, diseases, pesticides, habitat destruction, and climate change are suspected to contribute to this decline. In addition to their decreasing numbers, climate change has also caused mismatch between the timing of honeybee emergence and flower blooming. As plants and insects respond differently to changing temperatures, the timing of honeybee emergence and flower blooming can become misaligned, making it difficult for honeybees and plants to maintain their symbiotic relationship.

Without pollinating insects, agricultural production would decrease. In response, engineers and scientists have developed drones, or robotic bees, to replicate the work of bees. Interestingly, the term "drone" originally referred to male bees, and the creators of these small flying robots borrowed this word for their invention. The first robotic bees were developed by researchers at Harvard University. Their 2013 prototype could only fly or hover in the air and lacked additional functionality or remote-control capabilities. However, recent advancements have produced remotely controllable robotic bees that can flap their wings 120 times per second, ascend vertically, and hover around specific objects. In March 2018, the United States retail company Walmart filed a patent for robotic bees that can be remotely controlled and automatically identify pollen. This move appears to reflect Walmart's interest in agriculture or its commitment to providing consumers with a more stable supply of agricultural products. In 2021, researchers at the University of Maryland developed a palm-sized robotic bee for pollination (Fig. 7.2). Equipped with artificial intelligence, it can autonomously locate plants; avoid collisions with animals, trees, or other drones; and be controlled in groups by a single operator.

However, questions remain about how many different crops these robotic bees can pollinate, whether producing enough robotic bees to replace declining honeybee populations would impose a significant environmental burden, and the economic feasibility of their commercial use. These factors are still difficult to predict, and comprehensive economic analyses are currently lacking.

Fig. 7.2 A robotic bee recently developed by the University of Maryland. (Image provided by Nitin Sanket of Worcester Polytechnic Institute, USA)

Sexual Mimicry in Orchids Tricks Insects into Pollination

Some orchids exploit the reproductive instincts of male insects to transfer pollen to their pistils. For example, the sexually deceptive orchid (*Chiloglottis trapeziformis*), native to Australia, mimics the appearance of a female wasp with its flowers to attract male wasps (Fig. 7.3). When a male wasp attempts to mate with the flower, mistaking it for a female, pollen attaches to its body. If the wasp moves to another flower and tries to mate again, the pollen from the previous flower is transferred, completing pollination.

Researchers later discovered that the orchid also emits volatile compounds identical to the sex pheromones[3] secreted by female wasps [113], and these chemical signals were even more effective than the flower's shape. The pheromone was so powerful that male wasps attempted to copulate with a simple stick treated with it. With such an intricate and convincing lure, it seems no male wasp can resist the deception. The orchid depends entirely on this strategy for reproduction, apparently due to the coevolution of its unique flower shape and scent with the wasp. While the long history of evolution has produced many extraordinary forms and functions in plants and animals, the strategies of these orchids stretch the limits of human imagination.

Do these deceptive methods lead to a higher reproductive success rate compared to flowers that provide nectar as a reward to pollinators? Experimental results do not support this idea. In fact, fooled wasps quickly learn the deception and stop

Fig. 7.3 A male wasp (*Neozeleboria cryptoides*) attempting to mate with the flower of the sexually deceptive orchid (*Chiloglottis trapeziformis*), which closely resembles a female wasp. (Image provided by Rod Peakall of the Australian National University)

[3] A chemical signal used for communication among individuals of the same species, particularly in animals.

attempting copulation. Then, why did these orchids evolve such elaborate strategies? A few studies suggest that sexual deception may enhance the likelihood of outcrossing with other plants [99]. There is no doubt that curious scientists will continue to explore these and many other questions surrounding this remarkable plant-insect interaction.

Ants Coevolved with Plants

There are over 15,700 species of ants on Earth, found on every continent except Antarctica, with an estimated population of about 20 quadrillion individuals [114]. Ants inhabit diverse environments and consume a wide variety of foods. Among them, some ants cut leaves from plants, carry them back to their nests, and cultivate fungi on these leaves as food for their larvae (Fig. 7.4). Some ants even *cultivate* plants. In the tropical rainforests of Fiji in the South Pacific, ants collect the seeds of an epiphytic plant called *Squamellaria* and plant them in sunny spots beneath the bark of host trees. Once these epiphytic plants germinate and grow, the ants deposit their droppings into the plant's nutrient-absorbing structures, known as dormatia, providing the plants with essential nutrients. These carefully nurtured dormatia become homes for the ants, housing a single queen and up to 250,000 worker ants. This interaction can be considered a symbiotic relationship, as the ants disperse the plant's seeds, fertilize it, and protect it, while the plant offers the ants a comfortable habitat. Some even describe this as "ant agriculture," as the process of planting seeds, fertilizing, harvesting resources, and using the plants as living spaces mirrors human agricultural practices. Currently, 37 species of ants and about 200 plant species are known to engage in such symbiotic relationships.

Recently, a study suggested that ants' diverse ways of living across various environments may have been made possible through their coevolution with plants [89]. Both ants and flowering plants appeared on Earth around 140 million years ago and have since thrived as numerous species in a variety of habitats. Suspecting that their evolutionary success might be interconnected, Dr. Nelsen's team in

Fig. 7.4 Ants cutting and carrying plant leaves back to its nest. These ants use the leaf to cultivate fungi, which serve as food for its larvae. (Photo by Chris, via Flickr. Licensed under CC BY-ND 2.0)

Chicago began researching whether their evolution was linked. To find this connection, the team examined the temperature and precipitation data of habitats where 1400 species of ants currently live and compared this information with the ants' phylogenetic data. Phylogenetic information is akin to a family tree, using data from genetic sequences, morphological traits, and sometimes fossils to infer evolutionary relationships among organisms. When the researchers compared climate data with phylogenetic information, they found a strong correlation between the climate of an ant's habitat and its evolutionary lineage. For instance, ants whose relatives lived in dry regions were often found in similarly arid environments. This allowed the researchers to predict the climate of an ant's habitat based on its phylogenetic position, which they then extended to infer the habitats of ancient ant species. Applying the same approach to flowering plants, they compared the results with those of the ants. Their research concluded that plants and ants have maintained a close relationship over millions of years, helping each other survive and evolve into their diverse forms.

Let us take a closer look at the coevolutionary journey of plants and ants revealed by this research team. Sixty million years ago, ants primarily lived underground. Around this time, plants began developing stomata, releasing water into the atmosphere, which increased humidity and led to the formation of rainforests. The high humidity encouraged some ants to move their nests from underground to the trees. As flowering plants gradually expanded their habitats into drier regions, certain ant species followed, moving into these new environments alongside the plants. This diversification of habitats broadened the range of ants' living environments. Plants further facilitated this relationship by attaching structures rich in lipids and proteins, called elaiosomes, to their seeds. These elaiosomes attracted ants to carry the seeds, inadvertently spreading them across vast areas. As a result, plants that formed symbiotic relationships with ants were able to disperse farther and diversify more rapidly than those that did not. Likewise, ants that partnered with these plants discovered new habitats and multiplied in greater numbers. This mutualistic relationship enabled both plants and ants to expand their ranges and flourish in diverse forms. Through this interdependence, they actively shaped each other's evolution.

The Battle Between Herbivorous Pests and Plants

It is estimated that there are around 6 million species of insects on Earth, and approximately 50% of them feed on plants. These insects can pose a significant threat to the survival of plants. Since these herbivorous insects require plants to grow and sustain themselves, the type and abundance of plants in a given area largely determine the diversity and population of the insects that can survive there. Of course, interactions are never one-sided, and insects also have a profound impact on the growth, development, and reproduction of plants.

Insects like termites and cockroaches, which decompose dead plants, play a crucial role in recycling various nutrients within ecosystems. Without these insects, as well as other decomposers like fungi and bacteria, dead plants would not break down, and the Earth would soon be covered with layers of fallen leaves and dead trees, making it difficult for young plants to grow [95].

Among herbivorous insects, some feed on a variety of plants, while others rely exclusively on a single plant species. At first glance, eating a wide range of plants might seem advantageous for survival, but that is not always the case, as plants produce defensive toxins. Insects that specialize in consuming a single plant species have evolved to neutralize one or two specific defense compounds produced by that plant, allowing them to feed without harm from the plant's toxins. Some insects even use the plant's toxins to protect themselves. By absorbing the toxins and safely storing them in their bodies, these insects deter predators from eating them. Thus, insects that feed on a single plant species often utilize plants with specific toxins and use those compounds to shield themselves from predators. On the other hand, insects that feed on multiple plant species face the challenge of detoxifying a wide array of toxic substances produced by different plants. In such cases, if a plant detects an insect's attack and increases its production of defensive toxins, the insect's growth may slow down.

Herbivorous insects can be categorized based on the parts of plants they consume and their feeding methods. These include insects that feed on leaves, insects that insert straw-like structures into the phloem to extract sap, insects that attack only the roots, insects that form galls to divert the plant's nutrients for consumption, and insects that create tunnel-like structures within leaves.

Among these, the most common are insects that consume entire leaves and those that suck phloem sap (Fig. 7.5). When insects chew on leaves, they destroy various structures within the leaf, causing the stored substances inside to mix together. This mixing can trigger chemical reactions that produce toxic compounds harmful to the insect, potentially causing it harm. In contrast, insects that insert straw-like structures into the phloem seem to experience less damage from such mechanisms.

Herbivorous insects cause significant harm to agriculture. It is estimated that 20–30% of global agricultural production is lost to pests that destroy crops [112], and forest resources also suffer considerable damage. There have been multiple instances where non-native insects introduced from other regions have killed native tree species, leading to forest destruction [42, 126].

Fig. 7.5 Various ways insects feed on plants. The one on the left is an insect caterpillar that consumes entire leaves. The ones on the right are aphids. They insert their stylets into the plant's phloem and suck the sap

In 2020, regions in Africa, Southwest Asia, the Middle East, and southern China suffered severe damage from locust swarms. These swarms devastated crops in the area, making agricultural production difficult. While solitary locusts are not a serious problem, when three or four gather, they secrete pheromones that prompt swarm formation. As they gather, locusts consume more plants and reproduce at an astonishing rate, forming massive swarms. These swarms indiscriminately consume all nearby vegetation. Once resources are depleted, they migrate to new areas, devastating crops and forest plants along the way. This is especially harmful to farmers along their migration routes, as locusts pose a severe threat to their means of making a living from farming. The situation was especially dire in 2020, a year when the world was already grappling with the COVID-19 pandemic. In the affected regions (e.g., in East Africa; Fig. 7.6), locust swarms devoured crops, leaving farmers with scarcely anything to harvest and plunging them into hunger. Currently, the primary method of controlling these swarms is aerial pesticide spraying. However, this approach has serious drawbacks, as it kills beneficial insects as well, leading to ecosystem disruption.

How do locusts form such large swarms? What is the chemical nature of the pheromone that induces them to gather? Could there be a way to prevent locust damage without relying on pesticide spraying?

According to the research by Dr. Wang and Dr. Kang's team in China, locusts secrete a pheromone called 4-vinylanisole, which triggers them to form swarms [41]. As the number of locusts increases, the amount of pheromone they release also increases, attracting even more locusts and amplifying the swarm. This pheromone is detected by the OR35 receptor located in the locusts' antennae, which drives swarm formation. Mutant locusts that do not express the *OR35* receptor showed no response to 4-vinylanisole, and these locusts did not form swarms. Locusts that do not respond to pheromones in the laboratory would likely not form swarms in natural environments either. One potential solution could be releasing these non-swarming locusts into the wild, where they might interbreed with native locusts, creating harmless locust populations that do not form swarms. Another approach might involve spraying a substance that inhibits the function of the OR35 receptor,

Fig. 7.6 Desert locusts swarming the Samburu National Reserve in Kenya. In 2020, East Africa experienced its worst desert locust infestation in decades

preventing locusts from detecting the pheromone and inducing them to live solitary lives rather than forming swarms. Scientists are exploring these and other innovative ideas to mitigate outbreaks of pest insects that pose threats to crops.

How Do Plants Detect Insects Feeding on Them?

If most plants were completely defenseless against pests like locust swarms and other insects, very few plants would remain on Earth. However, the vast expanse of green vegetation across the planet demonstrates that many plants can effectively resist pests or recover from their damage. While agricultural crops with weaker defenses often succumb to fast-reproducing pests like locusts, most plants in natural ecosystems have mechanisms to detect the presence of pests and employ strategies to suppress them.

How do plants detect insects feeding on them? When insects chew on plants, they break cell walls, releasing intracellular substances. Among the many substances released, some are recognized by the plant (Fig. 7.7), triggering its defense mechanisms [1]. For instance, oligogalacturonides released from damaged pectin[4] in the cell wall bind to the FERONIA receptor at the plant cell membrane. This binding activates signaling pathways to start a defense response. Another set of

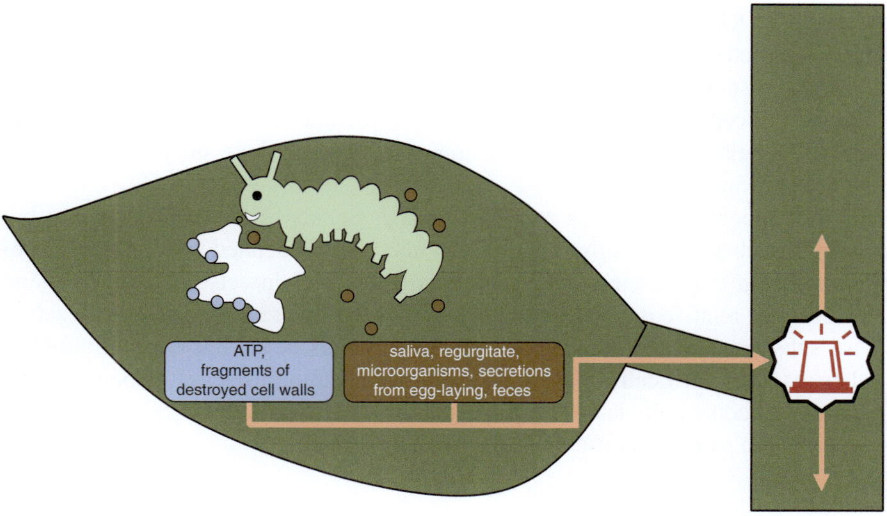

Fig. 7.7 How plants detect that they are being eaten by insects. Plants recognize various substances derived from both the plant itself and the insect during the feeding process. From the plant, these include ATP released from damaged cells and fragments of destroyed cell walls, such as pectin, oligogalacturonides, and cellulose fragments. From the insect, these include saliva, regurgitate, microorganisms associated with the insect, secretions from egg-laying, and feces

[4] A component of the plant cell wall, primarily composed of galacturonic acid.

experiments also supports the idea that plants perceive cell wall damage as evidence of pest feeding. Plants with defects in cellulose synthesis—a key component of the cell wall—recognize this as damage and activate their defense mechanisms accordingly [34].

In addition, insect feeding causes the release of ATP,[5] normally stored inside plant cells. The released ATP, now in the extracellular space, is recognized by neighboring cells, triggering processes that ultimately increase the synthesis of stress hormones such as jasmonic acid and ethylene. These hormones, in turn, activate the expression of various genes involved in the production of defense compounds against insect herbivores.

Researchers conducted experiments to identify the receptor that recognizes ATP released from leaves damaged by pests [21]. They observed that applying ATP to plant leaves increased cytoplasmic calcium levels and triggered a cascade of defense responses, including the expression of defense-related genes. The approach they chose was to search for mutant plants that did not show an increase in calcium levels upon ATP application. The researchers tested calcium levels in 50,000 mutant *Arabidopsis thaliana* plants after applying ATP and found two plants that did not respond. They discovered that these nonresponsive plants shared a single gene that differed from the wild type. They named this gene *DORN1* (*DOES NOT RESPOND TO NUCLEOTIDES 1*) and conducted follow-up experiments to test whether DORN1 was indeed the ATP receptor. An ATP receptor is expected to bind strongly to ATP, and DORN1 did so, consistent with the hypothesis. When they reintroduced the normal *DORN1* gene into the mutant plants that did not respond to ATP, the plants regained their responsiveness. Additionally, overexpressing *DORN1* enhanced the plants' responsiveness to wound signals. These findings established that DORN1 is the ATP receptor. Interestingly, DORN1 is structurally distinct from ATP receptors in animals, highlighting a molecular difference in how plants and animals respond to wounding.

Plants appear to recognize not only damage signals from wounded tissues but also cues such as insect saliva, regurgitate, microbes associated with insects, secretions released during egg-laying, and even feces from insects chewing on them. However, the relative importance of each of these factors remains unclear. If each of these substances were individually applied to plants at natural concentrations—rather than unrealistically high levels—and the resulting defense responses were measured, it could help determine the extent to which each substance contributes to triggering the plant's defense mechanisms.

---

[5] An abbreviation for adenosine triphosphate (ATP), an organic compound that provides energy for various biological activities within the cell. It functions like energy currency inside the cell.

Methods Plants Use to Defend Against Insects

What happens after a plant detects an insect attack? Plants employ various strategies to prevent insects from feeding on them, such as creating physical barriers, secreting chemicals harmful to insects, or producing defense proteins. Even if parts of the plant are consumed, the plant defense system can hinder the insect's digestion, impair its growth, and alter its physiology, ultimately preventing it from reproducing.

Plant surfaces are covered by *physical barriers* such as the cuticle[6] layer, trichomes,[7] epidermis, and bark, which deter insects from feeding on them. Dense hairs or sticky substances on the surface can hinder the movement of small insects. Additionally, hardened callus[8] tissue that forms after a wound heals serves as a tough barrier that is difficult for insects to penetrate. When plants detect an insect attack, they reinforce these physical defenses to make them even more robust.

Plants also employ *biochemical defense mechanisms*. For example, when a plant is damaged by an insect, chloroplasts inside the plant cells release a substance called linolenic acid. This fatty acid comes from the breakdown of the chloroplast's own membranes. Linolenic acid is then quickly turned into jasmonic acid, a hormone that helps the plant defend itself. In the plant *Arabidopsis thaliana*, jasmonic acid can be made within 30 seconds after the injury. It joins with another molecule called isoleucine and travels into the cell's nucleus, where it changes the activity of certain genes, leading to the production of proteins that help the plant fight off insect attacks.

Some plants synthesize peptides as danger signals and secrete them into the space between cell walls when insects feed on them. These peptides act as alarms, signaling the presence of an insect attack. For example, systemin in tomatoes is an 18-amino-acid peptide that spreads throughout the plant after it is wounded. Systemin induces the synthesis of proteinase inhibitors, which hinder the insect's ability to digest tomato leaves.

Because insects can move around, it is not enough for a plant to defend just the area that has been injured by the insects. If only the damaged part responds, the insect might just crawl to another spot and keep eating. That is why plants need to activate their defenses across their whole body—not just where they were hurt. Jasmonic acid, a key plant defense hormone, helps make this happen. It travels through the plant's internal "pipe system" (called the vascular system) and tells other parts of the plant to get ready. As it moves, more jasmonic acid is made in those areas, even in parts that have not been damaged yet [68]. This boosts the plant's defenses everywhere, so it is better prepared to stop the insect in its tracks. As jasmonic acid moves through the vascular system, more of it is synthesized in

---

[6] A lipid-based polymer layer covering the outer surface of aerial plant parts. Made mainly of cutin and wax, it prevents water loss, protects against microbes and UV light, and makes the surface water-repellent—causing droplets to bead up and roll off.

[7] A trichome is a hair-like or spine-like outgrowth found on the surface of plants.

[8] A mass of unorganized, proliferating plant cells that forms over wounded or cut surfaces, often serving to protect the area and facilitate regeneration.

vascular cells and in undamaged tissues upon its arrival. In this way, even in tissues that are not directly injured, jasmonic acid levels increase and trigger the expression of various defense-related genes, initiating a robust defense response.

The critical role of jasmonic acid in defending undamaged tissues from insects was revealed in the experiment depicted in Fig. 7.8. Researchers grafted the roots of mutant tobacco plants impaired in jasmonic acid signaling onto the shoots of wild-type tobacco plants. As a control, they grafted wild-type tobacco roots onto wild-type tobacco shoots to check whether the grafting process itself influenced the results and, if so, to compensate for it. They then wounded the leaves of these mixed and control plants and measured the levels of nicotine in neighboring, unwounded leaves. As a result, tobacco plants with disrupted jasmonic acid signaling in their roots accumulated less nicotine in unwounded neighboring leaves compared to control plants, rendering them more susceptible to attacks of leaf-eating insects [38]. This experiment demonstrates that when insects attack tobacco leaves, jasmonic acid activity in the roots is necessary for accumulating the insect-repelling nicotine in the neighboring leaves.

Jasmonic acid spreads throughout the plant via the phloem—a kind of internal transport system—but moves at a rate of only 1–5 cm per hour, too slow to trigger a rapid response [110]. Since plants begin systemic defense responses within about

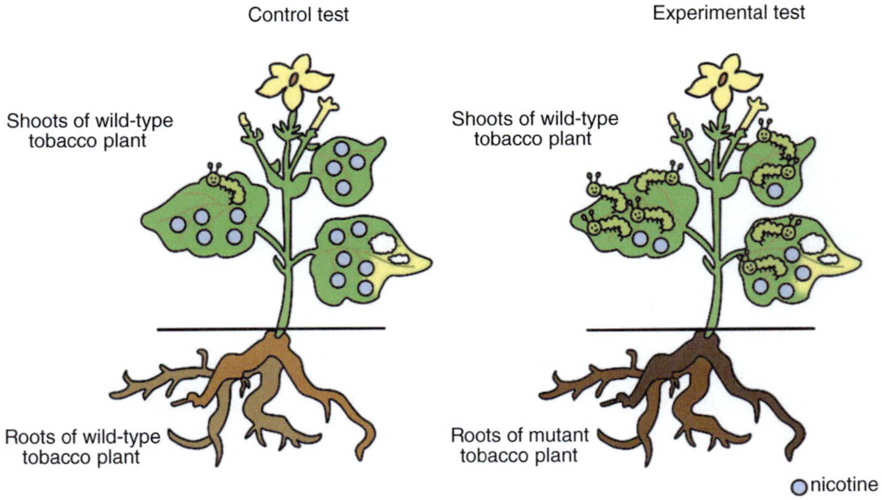

Fig. 7.8 An experiment demonstrating the importance of jasmonic acid signaling in the roots for systemic defense activation. The roots of a mutant tobacco plant impaired in jasmonic acid signaling were grafted onto the stem of a normal (wild type) tobacco plant (right), and as a control, the roots of a wild-type tobacco plant were grafted onto the stem of another wild-type tobacco plant (left). One leaf was wounded, and the accumulation of nicotine in neighboring leaves and the degree of insect attraction were measured. In the plant with jasmonic acid signaling-impaired roots (right), less nicotine accumulated in the leaves, making it more vulnerable to pests. These results indicate that even when the roots are not directly attacked by pests, jasmonic acid signaling in the roots plays a crucial role in activating systemic defense mechanisms

15 minutes of wounding, another, yet unidentified, signal likely spreads through the plant first, subsequently inducing jasmonic acid synthesis in more remote locations. One hypothesis suggests this signal could be electrical, while another points to volatile compounds such as methyl jasmonate, indole, terpenes, or green leaf volatiles. Methyl jasmonate, in particular, is highly volatile, easily penetrates cell membranes, and can move to other leaves of the same plant or neighboring plants, making it a potential mobile wound signal.

Volatile Compounds That Defend Plants Against Insects

Plants attacked by insects release distinctive volatile organic compounds into the air, which attract the predators of insects. For instance, when cabbage white butterfly larvae feed on plants in the Brassicaceae family, the wounded areas emit a variety of green leaf volatiles. These scents attract a small wasp species known as the white butterfly parasite (*Cotesia glomerata*), which lays its eggs inside the body of the butterfly larvae. The wasp larvae hatch and consume the caterpillar from within, preventing it from reaching adulthood and ultimately killing it (Fig. 7.9). But how can we be certain that the wasps are specifically attracted by the green leaf volatiles and not responding to some other signal? Experimental approaches could help clarify this intriguing question.

Fig. 7.9 Some volatile compounds secreted by plants attract the natural enemies of insects. When the larvae of the cabbage white butterfly (*Pieris rapae*) feed on the leaves of Brassicaceae plants, the plants release green leaf volatiles into the air. These scents attract the parasitoid wasp (*Cotesia glomerata*) which lay their eggs inside the bodies of the cabbage white butterfly larvae. The wasp larvae hatch and feed on the caterpillars, disrupting their maturation into adults and thereby inhibiting leaf consumption by the caterpillars

Dr. Takabayashi's research team in Japan addressed this question using genetic engineering technology [118]. Their hypothesis was that wasps are attracted to the scent of green leaf volatiles emitted by damaged leaves (Fig. 7.9). If this hypothesis were correct, they predicted that genetically modifying plants to suppress the production of green leaf volatiles would result in fewer wasps being attracted. To test this, they reduced the expression of genes encoding the enzymes responsible for producing the green leaf volatiles ((E)-2-hexenal and (Z)-3-hexenyl acetate in *Arabidopsis thaliana*). When attacked by caterpillars, these genetically modified plants emitted less green leaf volatiles, leading to fewer wasps being attracted and more leaves being consumed by the larvae. Conversely, when these genes were over-expressed, the plants produced more green leaf volatiles, attracting more wasp predators and reducing the damage to the leaves caused by the larvae. These experimental results established that green leaf volatiles effectively attract the natural predators of herbivorous insects, providing the plants with a defense mechanism against pests.

Another set of experiments on this topic involved fava bean (*Vicia faba*) plants under attack by pea aphids (*Acyrthosiphon pisum*). In this case, the researchers investigated whether aphid-infested fava bean plants could signal neighboring plants to release green leaf volatiles that attract the wasps (*Aphidius ervi*) which parasitize the aphids. Research by Dr. Guerrieri's team revealed that aphid-infested fava beans release L-DOPA into the soil (Fig. 7.10), thereby triggering uninfested

Fig. 7.10 How fava bean plants call their neighbors for help when attacked by aphids. When aphids invade a plant, not only the leaves of the attacked plant but also the leaves of neighboring plants release green leaf volatiles. This phenomenon occurs because the roots of the infested fava bean plant secrete L-DOPA, which spreads to nearby plants. Neighboring plants that detect L-DOPA also emit green leaf volatiles. The green leaf volatiles increase the expression of defense-related genes and attract the natural enemies of aphids, thereby helping to protect the infested fava bean plant

neighboring plants to emit volatiles that attract these wasps [17]. Further, the researchers treated healthy plants with L-DOPA and observed that the treated plants emitted three types of volatile compounds that attracted wasps, even in the absence of an aphid attack. These experiments established the role of L-DOPA in spreading the attack signal.

Strategies to Induce Predatory Behavior Among Pests

Plants also employ methods to manipulate herbivorous insects into preying on each other. A research team led by Dr. Orrock observed that when the plant defense hormone methyl jasmonate (MJ) was applied to tomato leaves, larvae of the beet armyworm (*Spodoptera exigua*) were more likely to engage in cannibalism and consumed less of the leaves [94]. By the end of the experiment, untreated leaves were almost entirely devoured by the larvae, whereas a substantial portion of the MJ-treated leaves remained intact. This was because MJ-treated leaves produced digestion inhibitors, such as tannins and proteinase inhibitors, which reduced the nutritional value of the leaves for the insects. The decreased nutrient availability intensified competition among the larvae, ultimately driving them to cannibalize one another to supplement the nutrients needed for growth. When pests consume one another, their overall population decreases, and the amount of plant material consumed per insect also declines. This makes the strategy highly effective in controlling pest infestations.

Chemical Compounds Plants Produce to Defend Against Pests

Plants produce hundreds of thousands of chemical compounds, most of which serve as a means of self-defense. These defense chemicals are referred to as secondary metabolites in contrast to primary metabolites, which are essential for basic survival. In the case of *Arabidopsis thaliana*, a key secondary metabolite for insect defense is glucosinolate. The importance of glucosinolates in insect defense is demonstrated by experimental results showing that mutant *Arabidopsis thaliana* plants with reduced expression of glucosinolate biosynthesis genes are more vulnerable to attacks from various insects.

One of the secondary metabolites produced by tobacco plants is nicotine. When tobacco plants are genetically modified to stop producing nicotine, insects consume the leaves much more readily. Conversely, even leaves that cannot produce nicotine naturally become less appealing to insects when nicotine is artificially applied to them. Tannins are another important secondary metabolite. The astringent taste of unripe persimmons and black tea comes from tannins, which bind to proteins and hinder their digestion. As a result, insects that consume tannin-rich leaves struggle to grow properly. This is why changes in the tannin concentration of poplar leaves can alter the types of insects that inhabit the tree [135]. Interestingly, many plant-derived substances that are harmful in large quantities are enjoyed in small amounts

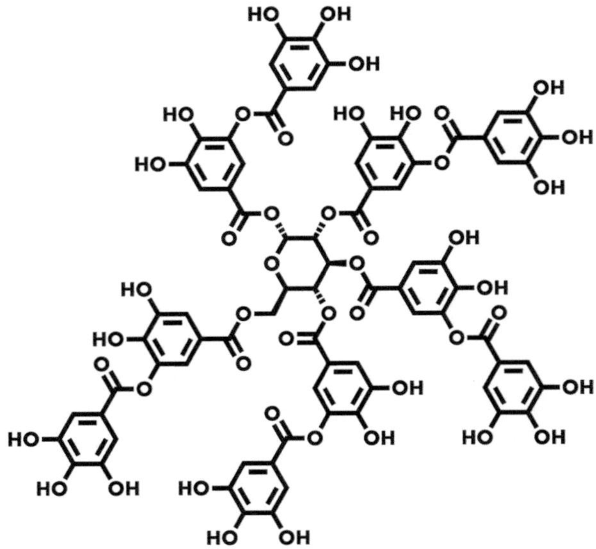

Fig. 7.11 One example of the chemical structure of tannins. The tannin molecule contains numerous hydroxyl groups (–OH) and double bonds, creating multiple regions that carry partial positive and negative charges within the molecule. This chemical structure allows tannins to bind effectively to proteins

by humans, such as the tannins in wine and tea. Additionally, many medicinal and psychoactive substances are derived from plant secondary metabolites.

Over the course of evolution, some insects have adapted to utilize secondary metabolites to their advantage. For example, certain species have developed specialized strategies to cope with glucosinolates—either by breaking them down or safely storing them—which enables them to preferentially feed on glucosinolate-containing plants without facing competition from other insects. Similarly, there are specific types of insects that tolerate tannins well and feed on plants rich in tannins (Fig. 7.11).

Methods for Unveiling the Molecular Mechanisms of Plant Defense Against Pests

When attempting to uncover which chemical substances plants use to inhibit or support other organisms, earlier researchers often narrowed their analysis to a few candidate compounds that were already known or suspected. If these predicted candidates were found to play no significant role, the research often stalled. However, with the advancement of precise metabolomic analysis technologies, such as UHPLC-HR-MS,[9] and IT tools capable of rapidly processing large amounts of experimental data, it is now possible to broaden the search of candidate compounds. A notable example of research using metabolomics is the study of the pest defense mechanisms in petunia plants [111].

[9] An abbreviation for ultra-high performance liquid chromatography high-resolution mass spectrometry, a chemical analysis technique capable of detecting even extremely low concentrations of chemical substances (For detailed explanation, see Appendix).

The authors of this study first discovered that a mutant petunia plant with reduced expression of the *PDR2* gene, a type of ABC transporter, was more susceptible to insect feeding compared to the wild type. This led them to hypothesize that the PDR2 protein might transport substances that help protect the plant from insect attacks. What substance does PDR2 transport to enhance the plant's resistance to insects? Since plants produce an enormous variety of chemical compounds for defense, identifying which of these are transported by PDR2 and inhibit insect growth was a challenging task. To overcome this, the researchers used UHPLC-HR-MS technology to comprehensively analyze the metabolites present in the leaves of both the mutant with reduced PDR2 expression and the wild type. Their analysis showed that several steroid compounds—highly toxic to insects—were found in lower amounts in the mutant than in the wild type. Based on this, the researchers speculated that these compounds could be the cargo transported by the PDR2 protein. Further experiments are needed to test this idea.

While the previous example focused on analyzing metabolites, recent advances in transcriptomics have enabled the comprehensive analysis of the expression of all genes in an organism. Using transcriptomic methods, researchers analyzed the changes in gene expression when insects chewed on leaves. The results revealed alterations in the expression of genes involved in processes such as ROS production and response, calcium signaling, cell wall reinforcement, secondary metabolite synthesis, plant hormone synthesis, and the regulation of transcription factors. An interesting finding from these diverse transcriptomic analyses is that the types of genes whose expression changes in response to leaf-chewing insects are remarkably similar across different plant species. However, when piercing-sucking insects, which insert their mouthparts to extract sap, attack plants, the pattern of gene expression changes differs from that induced by leaf-chewing insects.

Plant Proteins That Defend Against Insects

Dr. Ryan's research team in the United States demonstrated that plants under insect attack produce proteinase inhibitor proteins to effectively resist the assault [59]. They found that when tomato plants are damaged by insects or suffer mechanical injury, they produce proteinase inhibitors not only at the wounded site but also in other parts of the plant. These inhibitor proteins suppress the activity of digestive enzymes like trypsin and chymotrypsin in the insect's gut, thereby hindering the insect's growth.

In plants such as tomatoes and tobacco, enzymes that degrade specific amino acids—such as threonine and arginine—limit the availability of these essential nutrients required for insect growth [19, 40]. When these plant enzymes degrade the amino acids in the leaves, insects feeding on them experience difficulties in growth and reproduction. Thus, the enzymes contribute to pest control. This finding was confirmed through the following experiments: insects feeding on leaves from plants with reduced expression of these amino acid-degrading enzymes showed improved reproduction, whereas their growth and reproduction were inhibited when feeding on leaves from plants overexpressing these enzymes.

When aphids penetrate the phloem with their stylets to extract sap, plants produce specialized proteins that block the phloem or strengthen the cell wall to prevent the stylet from reaching the phloem. In *Arabidopsis thaliana*, a mutation in the *SLI1* (*SIEVE ELEMENT-LINING CHAPERONE1*) gene enhanced aphids feeding on phloem sap. The SLI1 protein is located on the outer walls of the phloem and can act like a clamp, tightening around the phloem and restricting the phloem sap flow. When aphids were introduced to both wild-type plants and *SLI1*-mutant plants (i.e., without a functioning SLI1), they fed for significantly longer and consumed more sap from the mutants [64].

Camouflage Strategies to Defend Against Insects

Some plants in the genus *Passiflora* have leaf protrusions that resemble orderly clusters of insect eggs (Fig. 7.12). These structures mimic the appearance of eggs laid by insects. When other insects encounter these protrusions, they appear to

Fig. 7.12 Egg-like protrusions on the leaves of *Passiflora* plants. These structures appear to be designed to prevent the longwing butterfly from laying eggs on the leaves. The butterfly is "tricked into thinking" that another butterfly has already laid eggs there and avoids the competition by seeking out other leaves. (Image provided by Deborah Austin of Grassy Knoll Exotic Plants (URL: https://gkplants.com/))

interpret the leaf as already occupied by another insect's eggs and avoid laying their own eggs on it.

Bladder Cells of Halophytes Prevent Insect Intrusion

The trichome cells on the leaves of quinoa, a halophyte that thrives in high-salinity environments, transform into large, balloon-like structures called bladder cells that cover the leaf surface. These bladder cells are spherical and typically 100 times larger than other leaf cells. About half of all halophytes possess such bladder cells. Since halophytes can survive in extremely high salt concentrations, up to 200 mM, many researchers have speculated that bladder cells store salt and water. There has even been hope that developing similar cells to cover the surfaces of ordinary plants could enhance their salt tolerance.

However, recent findings by Prof. Palmgren's research team in Denmark, presented in a paper and at the International Workshop on Plant Membrane Biology (IWPMB) conference in March 2023, suggest that the function of bladder cells is entirely different from previous assumptions [85]. To investigate whether bladder cells store sodium, the primary component of salt, the team isolated bladder cells and measured their ion concentrations. Surprisingly, they found that the sodium content inside bladder cells was actually lower than that of other cells, but the concentration of potassium ions was higher. To explore whether bladder cells might contribute to salt tolerance in another way, the researchers isolated a quinoa mutant lacking bladder cells and compared it with the wild type. The mutant plants exhibited salt tolerance comparable to that of the wild type. This unexpected result made it clear that bladder cells do not contribute to quinoa's salt tolerance. It was a surprising finding.

So why do these halophytes densely cover their leaf surfaces with bladder cells? To answer this question, the researchers conducted various experiments, such as exposing leaves to ultraviolet light or subjecting them to wind, but they were not able to find an explanation. As the research dragged on, the growth chambers became overcrowded with pots of quinoa plants, leaving no room for newly planted ones. To free up space, they moved some older pots to the hallway. Some days later, they observed that thrips were swarming only on the mutant plants, not on the wild type. This phenotype, which had gone unnoticed in the pest-free environment of the growth chambers, was revealed in the hallway. Seeking help from entomology experts, the team conducted further tests with various insect species. They found that larvae of the beet armyworm (*Spodoptera exigua*) also preferred and fed significantly more on the leaves of mutant quinoa plants lacking bladder cells compared to the leaves of wild-type plants covered with bladder cells (Fig. 7.13). Furthermore, bladder cells were found to be rich in toxic oxalic acid. This led to the conclusion that bladder cells serve as a combined physical and chemical barrier that prevents various pests from invading the leaves. They represent one of the remarkably diverse defense strategies plants use against pests.

Fig. 7.13 Quinoa mutant plants lacking bladder cells are less effective at defending against beet armyworm larvae. (**a**) Shows beet armyworm larvae after feeding on either wild-type (WT) quinoa plants with surfaces covered by bladder cells (left column) or mutant quinoa plants without bladder cells (right column) for 9 days. The larvae that fed on the mutant plants survived better and grew larger. The scale bar indicates 1 cm. (**b**, **c**) Show the appearance of plants fed on by larvae for 9 days. The youngest leaves damaged by the larvae are marked with arrows. Compared to wild-type plants (**b**), mutant plants (**c**) show significantly more damage, especially severe in younger leaves. (Image provided by Michael Broberg Palmgren of University of Copenhagen, Denmark)

Mimosa Plants Move to Repel Insects

The mimosa plant folds its entire leaves within seconds when it is touched or injured. According to a research by Dr. Hasebe's team at Japan's National Institute for Basic Biology, when mimosa leaves receive external stimuli, calcium ion concentrations change, and electrical signals sequentially spread throughout the leaf, causing it to fold [43]. Why does the mimosa fold its leaves in response to external stimuli? The researchers hypothesized that this behavior helps the plant repel its predators. They created a mutant mimosa that could not fold its leaves and compared it to the wild type. Interestingly, locusts, natural predators of mimosa, could not cling tightly to the folded leaves of wild-type mimosa and spent less time on them. In contrast, on the mutant plants that could not fold their leaves, the locusts stayed for long periods and fed on the leaves. This result demonstrates that mimosa plants quickly detect the physical stimulus caused by pests landing on them and fold their leaves as a defense mechanism to repel the pests.

This research made me appreciate Japan's remarkable emphasis on fostering basic science. Mimosa is not an agriculturally valuable plant, making it difficult to justify research on it for developing improved crops. Furthermore, as a legume, mimosa is challenging to mutate, requiring substantial time and funding for such studies. Despite these obstacles, this research was possible, likely due to the strong tradition in Japanese academia of valuing fundamental scientific exploration.

Mechanisms of Plant Responses to Insect Eggs

Female herbivorous insects are adept at recognizing chemical cues from their host plants, carefully laying their eggs on leaves, petioles, stems, or bark. These eggs are usually coated with substances secreted by the female, sticking to the plant like glue. It was previously assumed that because plants cannot move, they were defenseless when insects laid eggs on them, leaving them vulnerable to being consumed by larvae once the eggs hatched. However, recent studies have revealed that plants can recognize insect eggs and initiate defense responses against them.

When insect eggs are placed on a plant, the plant initiates various defense responses, including the production of ROS similar to a hypersensitive response, necrosis of plant tissue around the eggs, and the release of volatile compounds that attract natural enemies of the insect. For example, in *Arabidopsis thaliana*, when insect eggs are laid on its leaves, the plant generates ROS that kill the surrounding tissue, forms a hard callose barrier, and regulates the expression of hundreds of genes to mount a defense response. These responses are similar to the defenses triggered during pathogen invasion, with many genes induced through SA signaling [76, 79].

A research team led by Dr. Reymond in Switzerland found that when cabbage butterflies (*Pieris brassicae*) laid eggs on *Arabidopsis thaliana* leaves, SA accumulated not only at the site of oviposition but also in neighboring leaves [3]. Their study showed that the plant's response to insect eggs closely resembles its immune reaction to pathogen invasion (discussed in Part II). A particularly intriguing aspect of this SA-mediated response against insect eggs is that it also confers protection against bacterial and fungal infections. This systemic SA-mediated defense may act as a preemptive strategy, preparing the plant for potential pathogen entry through wounds created by hatching larvae.

How Much Should Plants Invest in Defending Against Insects?

Defense against insects is important, but if plants allocate too many resources to it, they may not have enough left for growth. When a plant invests heavily in defense but no enemies appear, those resources are wasted. In such cases, the plant may fall behind neighboring plants that focus more on growth. The question then arises:

when and to what extent should plants prepare defenses to optimize growth and reproduction?

A few years ago in April, insects swarmed the walnut tree in my yard, devouring all its green leaves within 3 days. The tree was stripped bare, resembling a winter tree with only branches left. It turned out that the neighborhood collectively sprayed pesticide in their gardens, but we were unable to do so at that time, and it seemed that all the insects in the area migrated to our tree as a result. After completely consuming the leaves, the insects disappeared. Remarkably, the tree quickly regrew its leaves and fully recovered its former appearance. By that autumn, it even produced more fully developed walnuts than the previous year. This led me to suspect that, rather than preemptively investing many resources in anti-insect defense, allowing leaves to be occasionally eaten and regrowing them might be a viable strategy for plants to thrive among competitors. In other words, foregoing defense and focusing on growth might be an effective strategy for competition in some situations.

Supporting this idea, there are reports on the disadvantages of overly intense defense responses. When plants are heavily consumed by insects, those that produce high levels of the defense hormone jasmonic acid recover more slowly than those with lower levels of the hormone. Scientists speculate that this may be because jasmonic acid interferes with gibberellins, the hormones responsible for promoting plant growth. While jasmonic acid is essential for anti-insect defense, excessive levels of this hormone can hinder a plant's ability to return to growth mode. This suggests that maintaining optimal jasmonic acid levels is crucial for plants to balance defense and recovery.

How Insects Overcome Plant Defenses

Insects use various strategies to overcome plant defenses. Some insects chemically detoxify the secondary metabolites synthesized and secreted by plants, while others break down plant cell walls to make it easier to insert their stylets into the phloem. Saliva secreted from the mouths of insects plays a crucial role in evading plant defense responses, too. What substances in insect saliva suppress plant defense responses? One key component of insect saliva is an enzyme that breaks down ATP. This enzyme degrades ATP released outside plant cells, weakening the ATP-triggered plant defense mechanisms. The importance of ATP-degrading enzymes is evident from several experiments. When artificially produced ATP-degrading enzymes are introduced to plants, the expression of defense-related genes, such as those encoding proteinase inhibitors and polyphenol oxidases, is suppressed. This strategy of herbivorous insects is similar to that of blood-sucking insects, whose saliva contains ATP-degrading enzymes that inhibit host defense mechanisms.

Some insects overcome plant defenses by utilizing symbiotic microorganisms. Beetle larvae secrete a large number of bacteria from their mouths. When plants detect bacterial flagella, they begin producing SA to defend against the bacteria. However, SA interferes with the action of jasmonic acid, which is crucial for

Fig. 7.14 Galls created on a beech leaf by the beech gall midge

defending against insects. In this way, insects use microorganisms to prompt plants to produce SA, thereby suppressing the jasmonic acid-mediated anti-insect defense mechanisms. The larvae of the moth known as the *spotted tentiform leafminer* (*Phyllonorycter blancardella*) also rely on their symbiotic bacteria. These bacteria produce the plant hormone cytokinin, which accumulates in the insect's salivary glands and is secreted onto the leaves as the larvae feed. This prevents leaf senescence, allowing the larvae to feed on healthy leaves and grow rapidly. Insects that exploit microorganisms to bypass plant defenses become significantly less effective at attacking plants when the symbiotic bacteria are eliminated using antibiotics.

Some insects create galls on leaves and hide inside them (Fig. 7.14). It is fascinating how they manage to bypass the plant's defense mechanisms to form and maintain these structures.

Applying Knowledge of Plant-Insect Relationships to Agriculture

Insects and pests that feed on plants harm agriculture by directly consuming plant tissues or by transmitting bacterial and viral pathogens to plants. To protect crops from such pests, farmers have traditionally relied heavily on broad-spectrum pesticides that kill multiple insect species at once. However, with the growing emphasis on environmentally friendly farming practices that minimize impact on nature, future agricultural methods will need to focus on pest control strategies that target specific harmful pests with minimal side effects, rather than using pesticides that indiscriminately kill a wide range of insects. Researchers have proposed various approaches to achieve this goal [32].

First, crops can be genetically engineered to produce bacterial toxin to protect themselves from pests. A representative example is crops expressing the *Bt* gene, widely used in modern agriculture.

Second, since the defense mechanisms plants use against insects and microorganisms share many similarities, enhancing a plant's immunity against insects by overexpressing proteins involved in anti-microbial defense is another approach.

Third, since plants naturally produce secondary metabolites that inhibit insect growth and reproduction, another approach is to enable plants to quickly and robustly activate genes for secondary metabolite synthesis in response to insect attacks.

Fourth, plants can be engineered to produce compounds that kill bacteria symbiotic with insects. Many insects cannot survive without their symbiotic bacteria, so eliminating these bacteria can effectively control insect populations. This approach is particularly effective against sap-feeding insects, which often lack immunity-related genes and rely on symbiotic bacteria to support their immunity.

Finally, for small-scale organic farming, utilizing the natural enemies of pests can also be a viable strategy. Identifying organisms that act as pest predators without harming crops and providing them with habitats or attracting them to the field can help manage pest populations. For instance, Rural Development Administration in South Korea recommended using entomopathogenic nematodes as a natural enemy to control pests that damage leafy vegetables like lettuce and radish greens [97]. Specifically, these nematodes are effective against pests like brassica leaf beetles (*Phaedon brassicae*) and striped flea beetles (*Phyllotreta striolata*), when applied during the early stages of infestation.

However, no method can completely and permanently eliminate pests. The sheer number of pests and their rapid reproduction rate allow them to eventually overcome the strategies that plants or humans implement for pest control. Therefore, to effectively protect crops from pests, it is essential to employ multiple methods simultaneously. To develop pest-resistant plants, breeding through classical cross-pollination has been widely used—and continues to be used. However, this approach is time-consuming and limited in producing crops with multilayered defenses against pests. In contrast, genetic engineering enables the rapid expression of multiple genes, allowing for faster development of plants with these desired traits. By identifying various methods to protect crops from pests and expressing the corresponding genes in combination, it is possible to construct multilayered barriers, enhancing plant defenses against pests.

Carnivorous Plants that Prey on Insects

Carnivorous plants that prey on insects are typically found in nitrogen-deficient environments, such as acidic wetlands. In these habitats, the decomposition of dead organisms is limited, preventing the formation of nitrates or ammonium. To survive, these plants capture and digest small insects to obtain proteins. This adaptation likely evolved to help them thrive in nutrient-poor conditions.

Carnivorous plants (Fig. 7.15) employ both passive and active methods to capture insects. Passive methods include secreting sticky substances, as seen in the Cape sundew (*Drosera capensis*), which immobilizes insects that become trapped,

Fig. 7.15 Various carnivorous plants. Clockwise from the top left: Venus flytrap (*Dionaea muscipula*), *Sarracenia purpurea*, *Nepenthes gracilis*, and Cape sundew (*Drosera capensis*).

or utilizing slippery pitfall traps, like those of the pitcher plants (Nepenthaceae), where insects fall into a slick opening and cannot escape once inside. Active methods involve mechanisms like those of the Venus flytrap (*Dionaea muscipula*), which snaps shut when an insect walks into its trap, preventing it from escaping. The traps carnivorous plants use to catch insects are typically modified leaves. Because they derive nutrients from their prey, their root systems are far less developed compared to those of other plants.

Charles Darwin was so fascinated by the swift and powerful movements of the Venus flytrap that he described it as "the most wonderful plant in the world." In his 1875 book, *Insectivorous Plants*, Darwin reported that carnivorous plants use protein-digesting enzymes, such as pepsin, to break down the proteins of captured insects for absorption. He concluded that this feeding behavior contributes to the plant's growth. Later researchers confirmed that carnivorous plants benefit from catching insects, as it promotes their growth and increases their reproductive success [128].

In the twenty-first century, Dr. Hedrich's group in Germany conducted detailed studies on the mechanism by which the Venus flytrap captures insects [49]. The leaves of the Venus flytrap have evolved into traps that close when touched, allowing the plant to detect the movement of insects on its surface. When an insect first touches one of the trap's sensory hairs, an electrical signal is generated. This initial signal only triggers a preparatory phase and does not yet close the trap. However, if a second electrical signal occurs within 30 seconds of the first, caused by further insect movement, the trap closes.

If the insect continues struggling to escape, its movement generates successive electrical signals, activating jasmonic acid signaling and causing the trap to tighten. When the electrical signal reaches five or more pulses, digestive enzymes are secreted from glands inside the trap to break down the captured insect (Fig. 7.16). These enzymes digest the insect's body into nutrients such as proteins, other

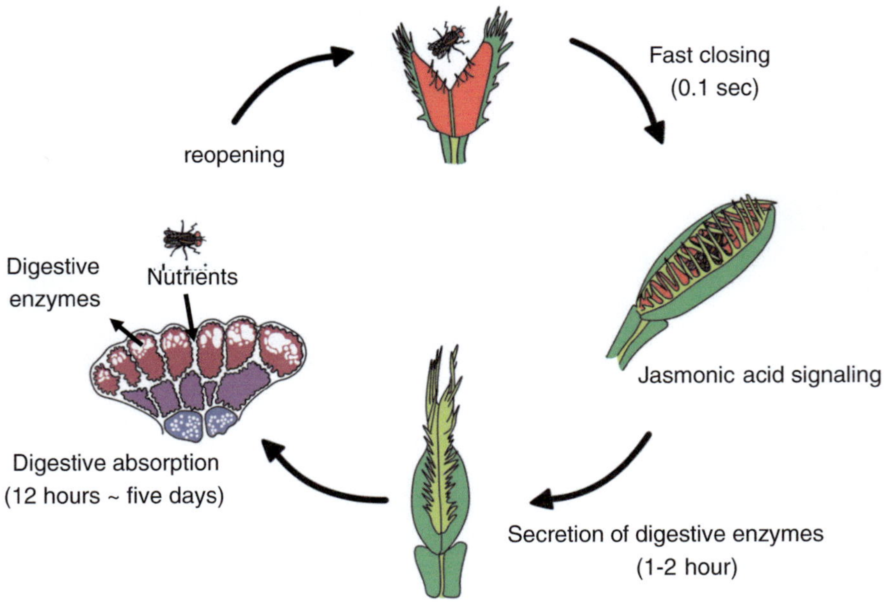

Fig. 7.16 How the Venus flytrap captures insects. When an insect passes by and touches the sensory hairs on the trap, an electrical signal is generated and spreads through the surrounding area of the trap. If another electrical signal is triggered within 30 seconds, the Venus flytrap's leaves snap shut, trapping the insect inside. Once the trap begins to close, the insect struggles harder to escape, which generates additional electrical signals. These signals cause the trap to close more tightly and stimulate the secretion of digestive enzymes. The enzymes break down the insect's body into nutrients that the plant can absorb. After the nutrients are absorbed, the trap reopens, ready to catch the next prey

nitrogenous compounds, potassium, sulfates, and phosphates, which the plant absorbs using transporter proteins. The transporter proteins that facilitate nutrient absorption in the trap are similar to the transporter proteins expressed in the roots of other plants. However, carbohydrates are not obtained from prey; like other plants, carnivorous plants produce them through photosynthesis. This is the reason why they thrive in sunny spots.

Carnivorous plants can detect the size of their prey through the intensity of the electrical signals generated by the insect's struggle and adjust proportionally amount of digestive enzymes secreted. Electrical signals exceeding five pulses also promote the expression of transporter genes, enabling the absorption of the nutrients released from the prey. Interestingly, even without capturing an insect, introducing jasmonic acid into the plant triggers the expression of nutrient transporter genes. Conversely, if the jasmonic acid signaling pathway is artificially suppressed, these transporter genes are not activated even when physical stimuli are applied. This suggests that physical stimuli are converted into electrical signals, which are subsequently transformed into the chemical signal jasmonic acid, activating the necessary genes.

After observing that the Venus flytraps close their traps after more than two electrical stimuli and secrete digestive enzymes after more than five stimuli, Dr. Hedrich remarked, "Plants can count. Plants can do arithmetic." If carnivorous plants were to close their traps or secrete digestive enzymes in response to mere wind or small debris like floating hairs, they would waste energy without gaining nutrients. By responding only to sufficiently strong stimuli, such as those caused by insects, the plants ensure efficiency in their nutrient acquisition. Dr. Hedrich playfully anthropomorphized carnivorous plants as "counting" to describe their selective and energy-efficient response mechanisms.

Recently, scientists have decoded the genomes of two carnivorous plants: the humped bladderwort (*Utricularia gibba*), which traps prey in tiny bladder-like sacs, and the corkscrew plant (*Genlisea aurea*), which uses modified, chlorophyll-free leaves to catch its prey. These plants have about the same number of genes as other flowering plants, and most of their genes seem to play similar roles, though many of them still have not been fully characterized. Genes involved in rapid response to external stimuli and movement, similar to those found in animal nervous systems, were not identified in these carnivorous plants. Moreover, there was no evidence of genes acquired from the animals they consume. This genome analysis suggest that carnivorous plants use genes common to all plants to create and operate their highly specialized traps, which is both fascinating and surprising.

The Origins and Endangered Status of Carnivorous Plants

To date, approximately 630 species of carnivorous plants have been identified. These plants belong to six different orders of flowering plants and seem to have evolved independently rather than diverging from a single common ancestor. Despite their separate evolutionary paths, the shapes of their traps and the methods they use to digest prey are remarkably similar. These carnivorous species, during long evolution to capture insects as a means of obtaining nutrients in nutrient-poor environments, eventually converged to develop similar forms and biochemical mechanisms. Carnivorous plants may have repurposed their defense-related genes for use in capturing and digesting insects, since the substances secreted by the Venus flytrap include a high proportion of hydrolytic enzymes and antimicrobial compounds, substances typically produced by plants to protect themselves from pathogens.

However, half of all carnivorous plant species are now endangered [56]. Habitat destruction caused by wetland drainage for agriculture, sensitivity to pollution, and overcollection by people are among the primary causes of their decline. While many are fascinated by these unique plants and want to keep them for personal observation, most people cannot provide the proper environment needed for their growth and reproduction, leading to their inevitable death shortly after collection. The disappearance of carnivorous plants from ecosystems would impact the survival of other species that depend on them. For example, frogs that use pitcher plants as shelters would struggle to survive if the habitats of pitcher plants were destroyed. Additionally, carnivorous plants play a role in controlling harmful pests, such as mosquitoes, flies, and *Schistosoma*, which cause diseases in humans. Without these plants, the prevalence of pest-related diseases could increase, causing suffering. Therefore, it is necessary to take protective measures to prevent their extinction.

8 The Relationship Between Plants and Vertebrates

Plants Feeding and Using Vertebrates

Plants, being stationary organisms, require methods to disperse their seeds. If seeds simply fall near the parent plant, germinate, and grow there, the parent plant and its seedlings would eventually compete for the same resources in the same location. Such intraspecific competition would make it difficult for an individual plant species to proliferate over wide areas.

To prevent this, many plants encase their seeds in enticing, fleshy fruit to encourage herbivores to eat them and disperse the seeds over greater distances. For instance, mistletoe uses birds, while durian relies on orangutans for seed dispersal.

Some plants disperse their seeds—encased singly in dry non-tempting fruits, achenes—by attaching them to animals' bodies. For example, the achenes produced by Spanish needles have hooks that allow them to cling to animals, enabling them to travel long distances (Fig. 8.1). Among such plants, the "bird-catching tree" exhibits a particularly fascinating phenomenon. Certain trees in the genus *Pisonia*, found on islands such as those in New Zealand and Hawaii, are locally known as "bird-catching trees." These trees do not capture birds to supplement nutrients, as they grow in nutrient-rich soils. Instead, living in the unique habitat of islands with limited means for seed dispersal, they have adopted a strategy of relying on birds that fly long distances and nest on other islands to spread their seeds. Since many seabirds do not eat fruit, *Pisonia* trees cannot use the typical method of encasing seeds in edible fruit. Instead, these trees produce seeds with hook-like structures coated in a sticky, glue-like substance that adheres to birds' feathers. Once attached, it is nearly impossible to remove the seeds. During the fruiting season, a bird that perches on a *Pisonia* tree may end up with a few seeds stuck to its feathers. When the bird flies to another island to nest in the next season, any feathers carrying seeds that fall to the ground allow the seeds to germinate and grow in a new location.

Y. Lee, B. Y. Choi, *Social Lives of Plants*,
https://doi.org/10.1007/978-981-96-9837-0_8

Fig. 8.1 Spanish needles dispersing seeds using animals. From left to right, clockwise: My dog Bokja examining Spanish needles (*Bidens alba*); seeds of Spanish needles attached to Bokja's fur; and the seeds removed from Bokja's fur and arranged in order

However, there is an occasional downside: unlucky birds with too many seeds stuck to their feathers cannot spread their wings, becoming immobile and unable to forage, which ultimately leads to starvation. It is this extreme side effect of the *Pisonia* tree's dispersal strategy that has earned it the name "bird-catching tree." The tree's strategy to expand its range from the limited environment of small islands has resulted in the accidental deaths of the birds.

Plant Defense Mechanisms Against Vertebrate Herbivores

Plant defense mechanisms against large herbivores can be broadly categorized into physical defenses and chemical defenses. Physical barriers that deter vertebrates from feeding on plants include sharp thorns and thick bark (Fig. 8.2). The chemical defense mechanisms plants use against herbivorous vertebrates are similar to those used against insects. This similarity arises because both insects and large vertebrate herbivores wound plant leaves, triggering comparable defense responses, such as wound-induced production of jasmonic acid, which induces the production of protease inhibitors and various secondary metabolites.

Fig. 8.2 Physical defenses of plants against herbivores. On the left: The prominent thorns of the prickly castor oil tree (*Kalopanax septemlobus*) help deter herbivores from feeding on its leaves. On the right: A pine tree with thick bark that protects it from vertebrate herbivores

The Impact of Vertebrate Herbivores on Plant Diversity

In grasslands, grazing animals feed on leaves of young trees, hindering the growth of young trees into mature ones. Herbaceous plants are also affected by the feeding activity of herbivores. For instance, forests in the northeastern United States, where white-tailed deer populations have been high for several decades, have seen declines in both the number and diversity of understory plants. The research team led by Dr. Pendergast studied this issue over 11 years and found that even after the deer were removed, plant diversity did not immediately rebound [100].

What was the cause? Was it due to a reduction in the number of grass seeds? Could plants unpalatable to deer became dominant, occupying all available space and reducing plant diversity? Researchers who investigated these questions were surprised to find that even after the deer population decreased, the population of grasses favored by deer did not increase. At the same time, the number of plants unpalatable to deer did not increase either—in fact, it slightly declined [44]. What had significantly changed in areas with high deer populations was the soil itself: the soil density had increased, decreasing its aeration. Researchers interpreted these observations to mean that the degradation of soil quality—likely caused by the long-term presence of large deer populations—contributed to the slow recovery of herbaceous plant diversity.

When large herbivores like elephants increase in number, the growth and reproduction of the trees they feed on become difficult, favoring the survival of plants equipped with defenses such as thorns or chemical deterrents. Elephants uproot, push over, strip bark, and consume sap from tall trees, hindering their growth or causing them to die. As a result, a decline in elephant populations often leads to an increase in tree density. On the other hand, feeding activity by elephants creates open spaces in forests, allowing light-demanding tree species to establish themselves.

Large herbivores face relatively little risk of predation and thus forage even in dense forests. In contrast, smaller herbivores tend to avoid thick forests where predators might be lurking. As a result, in environments with both large and small herbivores, plant distribution often forms a diverse mosaic pattern of grasses, tall trees, and shrubs, leading to high ecosystem diversity [8].

In certain environments, plant growth forms alternate over time. For instance, thorny tall trees can protect shrubs with leaves favored by herbivores. The thorns of these trees effectively shield smaller shrubs below. However, shrubs grow quickly, consuming large amounts of water and nutrients, and eventually outcompete the tall trees and take over the area. Since shrubs lack effective defenses against herbivores, they are eventually consumed and disappear. This creates open spaces that are recolonized by thorny tall trees, restarting the cycle.

As evident from the above examples, trees must protect themselves from herbivores to survive. Strategies trees use to endure herbivore attacks include secreting toxic substances, growing near plants that effectively deter herbivores, inhabiting steep or rocky areas inaccessible to herbivores, or living in locations where carnivores that prey on herbivores are present [8].

The Triangular Relationship Between Plants, Herbivores, and Carnivores

What would it look like if we included not only large herbivores and plants but also carnivores to see what kind of relationships exist among these three? Dr. Ford of Canada, in collaboration with research teams from Kenya and the United States, conducted studies on this topic in Africa [37]. Their research addressed questions such as whether carnivore activity is required for the sustained growth of plants in the presence of herbivores (i.e., whether carnivores promote plant growth by preying on herbivores and reducing their numbers) or whether plants can continue to grow since they independently defend themselves against herbivores through mechanisms like producing toxic substances or developing protective structures such as thorns.

To address these questions, the researchers observed impalas, a type of herbivorous antelope in Africa, along with their predators—leopards and wild dogs—and the *Acacia* trees that serve as a food source for the impalas. Impalas tended to avoid dense forests during times when leopards and wild dogs were active, instead forming groups in open spaces. This behavior is likely due to the fact that dense forests, with limited visibility, increase the risk of surprise attacks by predators that impalas cannot see approaching. However, there was also the possibility that impalas avoided forests because the trees in open spaces were more palatable. To clarify which explanation was correct, the researchers created five experimental plots, each covering 5000 square meters, where they removed all trees. This eliminated the tasty trees but also removed any risk of predators hiding for ambush, creating completely open areas. By tracking five groups of impalas using GPS over a period of 60 days, the researchers found that impalas spent 2–6 times more time in the newly cleared areas. Based on these findings, the researchers concluded that impalas select open spaces primarily to avoid the risk of encountering predators.

Next, the researchers investigated whether the feeding behavior of impalas influenced the habitat selection of *Acacia* species. In the study area, there were *Acacia* species with short thorns and others with long thorns. By tracking which species of *Acacia* leaves the impalas fed on, the researchers found that impalas ate 1.4 times more leaves from the bitter-tasting, short-thorned trees compared to the long-thorned ones. However, they could not conclusively attribute this preference solely to thorn length. To test this, they removed the thorns from long-thorned *Acacia* trees, after which impalas consumed nearly the same amount as they did from the short-thorned trees. Conversely, when artificial long thorns were attached to short-thorned *Acacia* trees, the impalas reduced their consumption to levels similar to that of the naturally long-thorned trees. The conclusion was that thorn length significantly influenced impala feeding preferences. In other words, the *Acacia* trees' strategy of allocating resources to thorn production effectively defended them against impalas.

But does this impala feeding behavior actually influence the distribution of *Acacia* species? To answer this question, the research team set up nine 10,000-square-meter experimental plots from 2009 to 2014 for comparison. In areas fenced off with electric wire to prevent impalas from feeding, the population of short-thorned *Acacia* trees increased by 233%. Conversely, in areas where impalas had unrestricted access to feed, the number of long-thorned *Acacia* trees was seven times higher than in the fenced-off plots. The short-thorned trees were heavily grazed by impalas, whereas the long-thorned trees were left untouched, allowing them to grow more abundantly.

To further test whether impalas feeding on short-thorned *Acacia* trees significantly influence the distribution of *Acacia* species, the research team randomly selected 108 plots, each 200 square meters in size, from open spaces and thickets in an African study area. In each plot, they recorded the number and type of *Acacia* trees present, as well as whether the plot was dense forest or open space. The results showed that short-thorned *Acacia* trees were more abundant in forests, while long-thorned trees were more prevalent in open spaces. Combining these findings, the researchers concluded that in forests with predators like leopards and wild dogs, impalas are scarce, reducing the risk of *Acacia* trees being grazed. As a result,

Fig. 8.3 The triangular relationship between acacia trees, impalas, and leopards. In densely wooded forests, the risk of leopard attacks is high, so impalas rarely venture near. In these forests, where impalas are absent, short-thorned acacia trees that prioritize growth have a competitive advantage over long-thorned acacia trees that invest in self-protection against impalas. Consequently, short-thorned acacias are more likely to survive the competition within the forest. Leopards, meanwhile, depend on large forest trees for cover, as their hunting success improves when they can ambush prey from behind vegetation or in shaded areas

Acacia species that allocate resources to growth rather than producing long thorns are more likely to thrive in such environments (Fig. 8.3).

Experiments described above conducted by Dr. Ford's team clearly demonstrated that the three-way relationship between plants, herbivores, and carnivores actually exists. In short, the distribution of the herbivore impalas influenced the morphology of the plants, while the distribution of the impalas was determined by the presence of carnivores. Lastly, the distribution of carnivores is, in turn, affected by the distribution of plants, as carnivores need to hide in the shade provided by large trees to successfully ambush their herbivorous prey.

Due to human activity, the populations of large carnivores like lions and leopards are steadily declining, causing substantial impacts on ecosystems. If leopards disappear, impalas will venture into densely forested areas. As a result, long-thorned *Acacia* species, which are more resistant to grazing, will thrive both in forests and grasslands. This could lead to plant distributions in forests and grasslands becoming increasingly similar.

Throughout Earth's history, up until relatively recently, every continent was home to very large herbivores, such as elephants. However, today, most of these large animals have disappeared. Between about 50,000 and 10,000 years ago, a mass extinction event occurred, wiping out many large mammals, and the primary cause might be human migration. The arrival of humans in regions such as Australia and the Americas coincides with the disappearance of their large herbivores, suggesting a causative connection. During this period, large animals were disproportionately affected compared to smaller ones, likely because humans hunted large animals more extensively. As mentioned earlier, large herbivores play a critical role in shaping the growth, distribution, and diversity of plants, meaning that changes in their populations can lead to shifts in entire ecosystems. Today, human activities continue to drive the rapid decline of large animals, and this trend is expected to accelerate ecological changes in the future [15].

A similar phenomenon is occurring in the oceans [5]. In the subtropical ecosystems off the western coast of Australia, sharks prey on marine mammals such as dugongs and sea turtles. As a result, dugongs and turtles tend to stay in shallow waters where the risk of predation by sharks is minimal. In these shallow areas, fast-growing and smaller seagrasses, which are less likely to be consumed by dugongs and turtles, are the dominant species. In contrast, in deeper waters where sharks are present, larger, slower-growing seagrasses thrive because dugongs and turtles avoid these areas. However, shark populations have been declining recently, primarily due to habitat destruction caused by human activity. A reduction in shark numbers allows dugongs and turtles to roam more freely and access a wider range of seagrass meadows, leading to an increase in their populations. The growth in dugong and turtle populations threatens the survival of the large, slow-growing seagrasses that were previously protected by sharks. Over time, these seagrasses may gradually disappear. Since seagrasses play a vital role in removing carbon dioxide from the atmosphere via photosynthesis, their decline could accelerate global warming. Indeed, analyses indicate that oceans with abundant shark populations store 60% more carbon than regions with fewer sharks. Similar effects of predators on carbon cycling are observed in coastal wetlands and mangrove forests. In short, human-induced declines in marine predators are reducing the carbon storage capacity of coastal ecosystems.

The Changing Relationship Between Plants and Herbivores

As atmospheric carbon dioxide levels rise, the nutrient content of plants decreases. The crops we consume will have more sugar but less protein and essential fatty acids, resulting in foods that taste sweet but are not as beneficial for the body. The reason for this is explained in detail in Chap. 12. This change in food quality will inevitably affect the feeding behaviors and reproduction of herbivorous animals. Additionally, elevated carbon dioxide levels alter the production of chemical compounds plants prepare for defense. In general, the levels of flavonoids, phenolic compounds, and tannins increase, while the levels of terpenoid compounds decrease.

The current rate of climate change is extraordinary, occurring more than 10 times faster than any period in the past 65 million years. Since 1960, the habitats of crop-eating pests and pathogens have shifted northward at an average rate of 2.7 km per year, consistent with changes in temperature. Notably, there is growing evidence suggesting that insects and herbivores are responding to climate change more rapidly than the plants they consume. For example, while mangrove trees along coastlines are moving northward at a rate of 1.3–4.5 km per year, the mangrove crabs that feed on them have been observed in the eastern Atlantic to migrate northward at a much faster rate of 6.2 km per year. Such discrepancies disrupt the previously stable relationships between these organisms. Indeed, research has documented changes in the behavior, habitat, diet, size, and reproduction of mangrove crabs [15].

Additionally, changes in ocean currents have led to temperate seaweeds in regions like the Mediterranean, Japan, the Gulf of Mexico, Australia, and South Africa being heavily grazed by newly introduced tropical fish. Similarly, in many subarctic regions, rising temperatures have led to reduced and less persistent winter snowfall, exposing trees and allowing herbivores to feed more extensively on aspen and other species. In contrast, in areas closer to the poles, increased rainfall has caused surface snow to harden, making it more difficult for herbivores to dig through the snow to find grass, thereby reducing their access to food during winter.

The rapid pace of climate change is expected to cause fluctuations in herbivore populations, leading to increased instability in entire ecosystems. Adapting to climate change is particularly challenging for large-bodied, long-lived animals compared to smaller insects. When large animals give birth during periods of insufficient food availability, their offspring struggle to survive, and their populations are the first to decline. This is one of the reasons why many large herbivores and carnivores are currently facing extinction. The rate of extinction for large animals today is comparable to the rapid loss of megafauna during the Ice Age. This destabilizes food chains, affecting a wide range of species. For example, in North America, the loss of predators that hunt white-tailed deer has led to a dramatic increase in the deer population. Their overgrazing has reduced plant diversity, necessitating policies to control deer numbers. However, there is a lack of consensus among people on how to address this issue.

In Part III, we explored how plants interact and coexist with animals within ecosystems. Animals influence the form and distribution of plants, while plants, in turn, affect animals. But which ultimately determines the characteristics and stability of an ecosystem? Is it plants or animals?

Researchers focused on animals are likely to emphasize how animal predation shapes plant life, while those who study plants may be more interested in how the presence of certain plants dictates which animals can inhabit the area. In this way, interpretations of nature depend on the observer's perspective and prior knowledge. From my perspective as a botanist, plants appear to be the foundational element in any ecosystem because they provide the primary food source for nearly all forms of life, including insects, herbivores, carnivores, and humans. The plants in a given region are adapted to their soil and climate, and the animals that can consume these plants inhabit the same area.

However, there is one animal with far greater influence than plants: humans. The exponential growth of the human population and their consumption of Earth's resources have profoundly affected countless plant and animal species, altered the climate, and destabilized entire ecosystems. In this context, human activities have become the most critical factor in shaping modern ecosystems. The future of these ecosystems will depend heavily on human actions, as humans hold the power to determine their fate.

Part IV
Plants and Humans

Today, the most dominant life form on Earth is humanity. The entire planet has been altered by human activity, and plants are no exception. Plants deemed useful by humans have become crops, cultivated on vast stretches of land, while wild plants have lost their habitats to these crops, resulting in reduced biodiversity.

Humanity's immense power began with agriculture, the cultivation and use of plants. Settling near cultivated lands, people gradually refined their farming techniques, eventually producing more than they needed for immediate survival. This agricultural surplus gave them the freedom to think beyond subsistence, pursue diverse activities, and accumulate knowledge across generations—ultimately shaping the sophisticated civilizations we see today. In a sense, it was plants that granted humans the leisure that enabled their influence to expand so profoundly.

In the relationship between plants and humans, agriculture is the most critical factor. According to the Food and Agriculture Organization of the United Nations (FAO), approximately half of Earth's habitable land is used for agriculture. In this part, we will explore how humans selected certain plants for cultivation as crops, improved them, and came to rely on them. We will also discuss the environmental changes caused by the extensive interactions between humans and plants.

Crops: Plants Cultivated by Humans

9

Changes That Occurred as Plants Became Crops

In the Paleolithic Age, humans gathered edible parts from wild plants and occasionally hunted animals to obtain calories and nutrients. About 10,000 years ago, humans began practicing agriculture, cultivating plants. Today, humanity cultivates approximately 150 plant species for consumption, and these cultivated plants are called crops. The relationship between humans and crops has continued to evolve through mutual dependence, to the point where over 8 billion people cannot survive without crops, and crops, in turn, struggle to survive in the wild without human cultivation and management.

Then, how have plants changed during the 10,000 years of agriculture? Modern crops differ greatly from their wild ancestors (Fig. 9.1). Compared to wild plants, crops have become sweeter, more visually appealing, less sour and bitter, and free from toxins, adopting traits preferred by humans. Seeds and fruits have grown larger, and yields have increased dramatically. The seeds now remain attached to the grain-bearing ears, making them easier for farmers to harvest. Many crops have also evolved a single, upright stem without side branches, allowing for denser planting and higher productivity. Some can now propagate asexually,[1] simplifying and

[1]A mode of reproduction in which an organism generates offspring without the involvement of reproductive cells, producing new individuals from parts of its own body. The offspring are genetically identical to the parent, enabling the preservation of desirable traits, such as those in high-quality crops. In contrast, sexual reproduction involves the formation and fusion of male and female reproductive cells.

Y. Lee, B. Y. Choi, *Social Lives of Plants*,
https://doi.org/10.1007/978-981-96-9837-0_9

Fig. 9.1 Differences between modern crops and their ancestral wild plants. Modern cultivated maize has developed larger seeds and lost its side branches compared to its ancestral plant, Teosinte (*Euchlaena mexicana*). Rice has become easier to harvest due to reduced seed shattering, in contrast to ancient rice, whose seeds naturally fell off the panicle. Sunflowers have lost their numerous side branches, growing a single tall stem that produces one large flower. The absence of side branches allows sunflowers to be planted more densely, significantly increasing harvest yields

speeding up cultivation, while others have reduced seed dormancy,[2] resulting in more uniform germination. Synchronized flowering and fruiting have further improved harvesting efficiency.

While these changes benefit farmers, they disadvantage plants in the wild. Bitterness and toxicity could deter pests and diseases; easily dispersed seeds could spread far and wide, increasing the chances of population growth in various locations. A profusion of branches could allow individual plants to capture more space and light, gaining a competitive edge over neighboring plants. Thus, the loss of traits essential for plant survival is an unnatural change, a result of humans selectively breeding plants to be easier to cultivate and eat [54, 84].

[2] A state in which a mature seed temporarily delays germination, even under favorable conditions. This pause helps the seed avoid germinating too soon or too close to the parent plant, reducing competition for resources with the parent and sibling seedlings. Dormancy increases the seedling's chances of surviving and growing in a better location or at a better time.

Why Pesticides Are Necessary in Agriculture

Over the 10,000 years since agriculture began, farmers have selectively cultivated grains and vegetables that yield abundant harvests and have mild, appealing flavors. The plants that farmers prefer to cultivate are those that produce nutritious and delicious seeds and fruits. However, insects and microorganisms also favor such plants. Crops, in particular, have weaker resistance to pests and diseases compared to wild plants. This is because crops have been bred to allocate most of their energy and resources to producing desirable traits, such as tasty fruits and seeds, leaving little for producing bitter defense compounds against bacteria, fungi, viruses, and pests. It is similar to managing a limited budget—spending heavily on one priority, such as social welfare, reduces resources for other areas such as national defense. In contrast, wild plants, which grow without human intervention, produce smaller seeds and less sweet, often bitter fruits. These plants, however, allocate more resources to producing defensive compounds, making them naturally more resistant to pests and diseases.

To protect crops from pests, farmers rely on pesticides. These chemicals are specifically designed to suppress or eliminate insects and pathogens that target crops while minimizing harm to humans, other animals, and the environment. Additionally, most pesticides are formulated to break down quickly after use. Developing safe pesticides that are harmful only to target organisms and harmless to humans or other animals requires a detailed understanding of the fundamental biochemical differences between bacteria, fungi, insects, and animals. This knowledge allows researchers to design chemicals that are both effective and safe. Despite the wide variety of pesticides currently available, there is a continuous need for new pesticides. This is because repeated use of a single pesticide can lead to the emergence of resistant pests and pathogens, making the pesticide ineffective. As a result, the development of new, innovative pesticides remains a critical area of agricultural research.

Why Fertilizers Are Necessary for Farming

Plants need nutrients to grow and survive. Through their roots, plants absorb nutrients from the soil, such as nitrogen, phosphates, potassium, calcium, magnesium, and iron. When humans harvest and consume parts of these plants, they are effectively removing nutrients from the soil, while providing their bodies with essential minerals for growth and good health. In natural ecosystems, when plants die and decompose, their nutrients are recycled back into the soil. However, in agricultural fields, crops are harvested and removed, disrupting the natural cycle. As this process is repeated over time, the soil gradually loses its nutrients. In this depleted state, the soil can still support plant growth, but the plants are weaker and less nutritious than they should be. If nutrient loss continues for many years, the soil may eventually become so exhausted that it can no longer support plant life. Even in such depleted soil, adding nutrient-rich materials like compost made from the remains or waste of plants and animals, or synthetic fertilizers, allows plants to grow once again.

In modern society, synthetic fertilizers are used to maximize agricultural production. The main components of synthetic fertilizers are nitrogen compounds, such as ammonium (NH_4^+) or nitrate (NO_3^-), which plants can readily use. Although nitrogen makes up 78% of the atmosphere, plants cannot directly utilize this abundant atmospheric nitrogen. This is because nitrogen gas (N_2) is chemically very stable and cannot be used to produce essential biomolecules like proteins and DNA. To address this, humans convert atmospheric nitrogen into ammonia (NH_3) for use in synthetic fertilizers. This process of transforming chemically stable nitrogen into ammonia requires extremely high temperatures and pressures, consuming a vast amount of energy. Each year, approximately 1–2% of the world's total energy consumption is used to produce nitrogen fertilizers—a staggering amount of energy.

The modern agricultural practice of monocropping[3] further exacerbates the need for fertilizers. Monocropping involves cultivating crops of the same genotype (genetic composition) to increase yield and maintain consistent quality. Semchenko's research team revealed why this practice demands more fertilizers through experiments with sweet vernal grass (*Anthoxanthum odoratum*) [116]. Their study tested whether genetic diversity within plant communities affects nutrient absorption. Groups of sweet vernal grass with high genetic diversity released a wider variety and greater quantity of root exudates compared to groups with less diversity. These exudates stimulated microbial activity around the roots, enriching the soil. Consequently, genetically diverse groups absorbed more nitrogen nutrients from the soil than less diverse groups (Fig. 9.2). Similarly, a review summarized evidence that planting crops together with multiple genotypes can increase yields [129]. However, from the perspective of farmers, managing and marketing mixed crops is far more complex and challenging compared to monocropping.

[3] An agricultural practice where the same crop species—often of a single, genetically uniform variety—is grown repeatedly on the same land year after year.

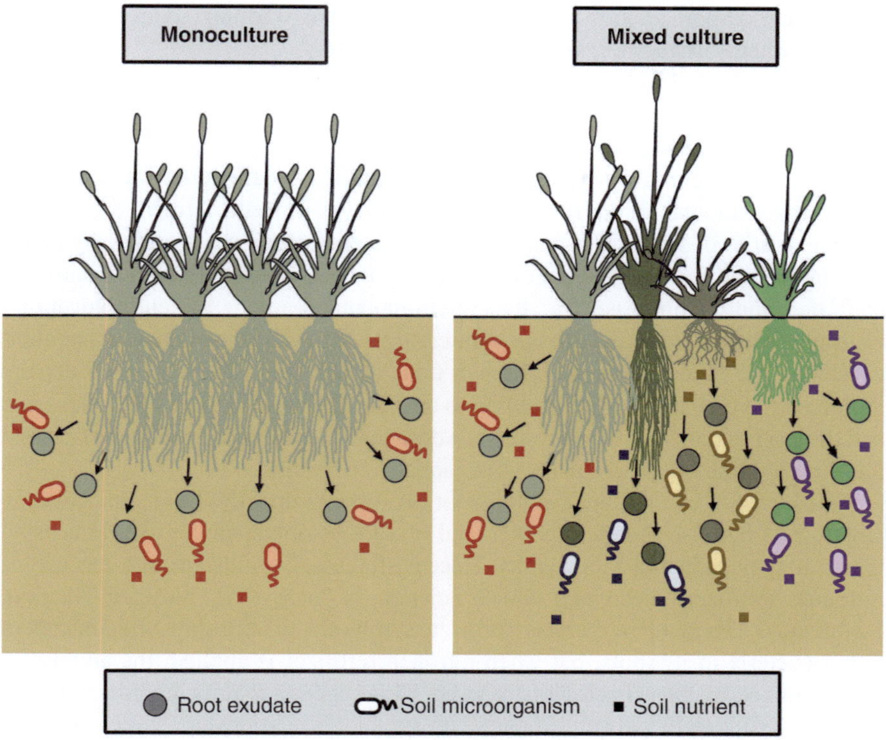

Fig. 9.2 Differences in soil fertility based on the genetic diversity of sweet vernal grasses. A monoculture of genetically identical sweet vernal grasses releases simple and limited root exudates, failing to enhance the activity of soil microorganisms. In contrast, a mixed stand of genetically diverse sweet vernal grasses produces a greater variety and quantity of root exudates, increasing the diversity and activity of surrounding soil microorganisms. This activity of various soil microorganisms enriches the soil and promotes plant growth

The Problems of Pesticides and Fertilizers

Why is the use of pesticides and fertilizers a problem? First, their production accelerates global warming. Manufacturing pesticides and fertilizers requires large amounts of energy, most of which comes from burning fossil fuels. This process emits carbon dioxide, a major cause of global warming. In addition, fertilizers lead to the production of nitrous oxide (N_2O), a greenhouse gas[4] that is 100 times more

[4] A gas that traps heat in the Earth's atmosphere by absorbing and re-emitting infrared radiation, thereby preventing heat energy from escaping into space. This contributes to the warming of the planet's surface.

potent than carbon dioxide in contributing to global warming. Nitrous oxide is released when soil microorganisms denitrify fertilizers.

Second, pesticides and fertilizers pollute the environment. Crops generally absorb only 30–50% of the fertilizers applied, with the remainder flowing into rivers, lakes, and oceans. This happens because crops cannot absorb and store large amounts of nutrients from fertilizer at once, and it is challenging for farmers to apply small amounts of fertilizer multiple times. As a result, farmers often apply large quantities of fertilizer at once. When fertilizers that are not absorbed by plants flow into water bodies, the concentrations of inorganic nutrients such as nitrogen and phosphorus increase in the environment, leading to eutrophication.[5] Eutrophication causes rapid growth of microalgae in the water, resulting in phenomena like algal blooms,[6] including green tides or red tides. These blooms deplete dissolved oxygen[7] in the water, leading to the suffocation of aquatic life such as fish and destroying underwater ecosystems. Additionally, some algae release toxins into the water, killing other organisms in the aquatic ecosystem.

Third, pesticides and fertilizers disrupt ecosystems directly. Pesticides not only target pests but also threaten the survival of various organisms in natural ecosystems, thereby reducing biodiversity. Recent studies also indicate that the use of nitrogen fertilizers decreases the efficiency of biological nitrogen fixation. Switchgrass (*Panicum virgatum*) forms a symbiotic relationship with nitrogen-fixing bacteria in its roots, but when nitrogen fertilizer is applied, the amount of nitrogen fixed by these bacteria is reduced. This suggests that the continued use of nitrogen fertilizers negatively impacts the symbiotic relationships between plants and bacteria, reducing the efficiency of biological nitrogen fixation and making agriculture increasingly dependent on chemical fertilizers [7].

The Reason Crops Require a Lot of Water

Without sufficient water, plants cannot maintain their structure. When water is scarce, plants wilt, unable to hold their fully extended leaf shapes, and they shrivel, much like a water balloon deflates and loses its form when the water is drained. If this condition persists, the plant eventually dies. However, if water is provided in time, the plant can recover its original structure (Fig. 9.3).

Plants cannot grow without sufficient water because plant cell growth depends on turgor pressure. This pressure is similar to the air pressure in an inflated tire. Turgor pressure is the result of water filling the cells. As water enters, plant cells expand and press increasingly on their walls, i.e., their turgor pressure increases. In

---

[5] A condition in which the concentrations of organic and inorganic nutrients become elevated in rivers, lakes, or oceans due to the inflow of chemical fertilizers or sewage.

[6] A rapid and excessive growth of algae in aquatic environments such as rivers, lakes, or oceans.

[7] The amount of oxygen dissolved in water, essential for the respiration of most aquatic organisms, including aerobic microbes, some plankton, fish, and shellfish. Adequate oxygen levels are critical for maintaining healthy aquatic ecosystems.

Fig. 9.3 A plant wilted due to lack of water recovers its shape after watering. A *Pelargonium* that had dried out over the weekend was watered ($t = 0$ h). After 3 hours, the leaves and flower stalks began to regain their turgor, and by 23 hours, the plant had fully recovered. These photos demonstrate that plants need sufficient water to maintain their structure

young plant tissues with thin extensible cell walls, turgor pressure drives permanent cell enlargement and tissue growth. When water is scarce, turgor pressure drops, preventing the plant cell from expanding and growing. Water is also crucial for nutrient transport in plants. When roots absorb enough water, it travels upward through the stem and leaves, distributing dissolved nutrients throughout the plant. Without adequate water absorption by the roots, these nutrients cannot be transported throughout the plant.

Because water is vital for plant structure, growth, and nutrient transport, it is indispensable for agriculture. Agriculture historically developed near rivers where water was readily available. The ancient civilizations of Egypt, Mesopotamia, the Indus Valley, and the Yellow River, all of which achieved remarkable agricultural advancements, were established around major rivers. Interestingly, the English word "rival," meaning competitor, is said to have originated from "river," referring to people competing for water resources. In 2018, two-thirds of the freshwater used globally was allocated to watering crops [102]. In drought-stricken regions, measures are sometimes taken to restrict water use for other purposes to ensure sufficient agricultural water supply.

The Problems of Irrigation and Irrigation Infrastructure

Since water is essential for plant survival, ensuring a steady water supply to plants is crucial for humans to obtain food from them. The proverb "drawing water to one's own field" describes selfish behavior but also underscores the urgency of watering crops to sustain a livelihood. The artificial supply of water to crops is called irrigation, and from ancient times to the present, irrigation infrastructure has been essential for agriculture to provide water in farmlands. In Korea, too, irrigation facilities have existed for centuries. One of the oldest examples is the Byeokgolje Reservoir, a large reservoir built in the Gimje area of the Baekje Kingdom period around the sixth century AD (Fig. 9.4).

However, constructing irrigation facilities can lead to environmental destruction. Actions like dredging riverbeds, leveling mountains, and erecting concrete structures to create irrigation flows inevitably alter natural ecosystems. These developments disrupt water flow, degrade water quality, trigger algal blooms, and destroy habitats, threatening biodiversity. In 2009, when the South Korean government launched the Four Major Rivers Restoration Project—claiming it would mitigate damage from droughts and floods—environmental groups voiced strong opposition. They argued that the project would not achieve its goals but instead devastate ecosystems and lead to environmental disasters. Conversely, the project has been credited with ensuring a stable supply of irrigation water for farmers and reducing flood damage. This controversy has been discussed in the international journal *Science* under the title "Restoration or devastation?" [91]. Only with time and more data will it be possible to objectively evaluate the effects and side effects of this project. What is clear now is that the divide between those who prioritize the necessity of irrigation for securing food and those who emphasize the importance of protecting the environment is unlikely to narrow.

Fig. 9.4 The sixth-century AD historic site of Byeokgolje Reservoir, the oldest irrigation reservoir in Korea. The site is located in Gimje, Jeollabuk-do. This water gate of the Byeokgolje reservoir, called Jangsaenggeo, is one of five water gates in this reservoir. It was used to supply water to the surrounding farmland

In addition to the issues caused by irrigation infrastructure, irrigation itself presents a challenge: increasing soil salinity. When water is applied to crops, much of it evaporates, but the dissolved salts remain in the soil. Over time, this process repeats, gradually raising soil salinity. Most crops cannot survive in high-salinity soils, as evidenced by the difficulty of growing crops in land reclaimed from the sea.

Genetic Changes in Plants That Improved Crops

10

The reason crops exhibit traits different from their ancestral plants is because there were changes in their genes. Until the latter half of the twentieth century, scientists did not know which genes in crops had changed. However, extensive research since then has identified more than 100 genes responsible for these differences.

How did genetic changes occur during the domestication process? In some cases, mutations absent in ancestral plants arose in crops, leading to new traits. In other cases, ancestral plants harbored multiple alleles[1] of a gene, but farmers favored specific traits associated with one of those alleles. Through the repeated cultivation of individuals exhibiting the preferred trait over many generations, farmers transformed crops to exhibit only the desired trait [29]. This process of selecting and cultivating plants with specific traits reduced the genetic diversity of domesticated plants.

A striking example of the extreme reduction in genetic diversity during the process of crop improvement is the navel orange (Fig. 10.1). Currently widely cultivated in countries such as the United States, Australia, and Brazil, the navel orange originated from a mutation first discovered in the 1820s at a monastery in Brazil. One branch of an orange tree in the monastery bore oranges with a distinct navel-like appendage. These oranges were highly desirable due to their sweetness, absence of seeds, and thick peel, which made them easy—even for children—to peel and eat. The mutation occurred in only a single branch of the tree, producing navel oranges. Subsequently, people propagated this mutated branch using grafting and cutting techniques, increasing the number of navel orange trees. Over time, this variety became one of the most widely produced orange cultivars in the world. Because all these trees were propagated asexually from the same original tree, they are genetically identical.

[1] A variant form of a gene found at the same position on a pair of chromosomes, inherited from each parent.

Y. Lee, B. Y. Choi, *Social Lives of Plants*,
https://doi.org/10.1007/978-981-96-9837-0_10

Fig. 10.1 The navel orange: an example of extremely reduced genetic diversity in crops. Millions of navel orange trees cultivated worldwide are propagated asexually, making them genetically identical to the original tree where the mutation first occurred. The navel-like structure of a navel orange is formed by the presence of a small secondary orange within the main fruit. This feature results from the incomplete development of a twin orange during its growth

Methods for Identifying Genes Altered During the Domestication Process

Traits in crops, such as yield, quality, and resistance to pests and diseases, are not simply categorized by the presence or absence of specific characteristics but rather exhibit a range of degrees. Such traits are referred to as quantitative traits. Quantitative traits are determined by the collective contributions of multiple genes in different regions on DNA strands, each playing a role in varying the extent of traits. For example, the size of seeds in the small weed *Arabidopsis thaliana* is determined by 11 quantitative trait loci (QTLs), each contributing incrementally to the trait.

In most cases, quantitative traits do not adhere to Mendelian inheritance,[2] where genes are transmitted independently to the next generation without interactions. Instead, these traits are influenced by complex interactions among multiple genetic variations and environmental factors, leading to a highly diverse range of trait expressions. Consequently, it is difficult to identify the genetic variations responsible for quantitative traits using simple crossing and breeding experiments. To uncover the genetic variations governing quantitative traits, researchers primarily use two methods: QTL mapping and genome-wide association studies (GWAS).

[2]A pattern of inheritance based on two fundamental principles proposed by Gregor Mendel. The law of segregation states that alleles for a trait separate into different gametes during the formation of reproductive cells. The law of independent assortment states that genes for different traits are inherited independently of one another without influencing each other. For example, in Mendel's pea plant experiments, the genes controlling seed shape and color were passed on separately. When traits follow these laws closely, it is possible to identify causal genes within just a few generations of crossbreeding.

QTL mapping is relatively straightforward in self-pollinating crops. First, a hybrid is created between the ancestral plant and the cultivated crop. The individuals in the first hybrid generation inherit one set of chromosomes from each parent—half from the ancestral plant and half from the crop. Some individuals may display traits from one parent, others may not, and some may exhibit intermediate traits that fall between the traits of the two parents. Subsequent generations are obtained through self-pollination, in which each hybrid plant is pollinated with its own pollen. Using this approach, the traits of the offspring become more varied and differentiated with each successive generation. This diversity arises from differences in the DNA sequences within the chromosomes of the offspring. The differences result from the process of meiosis,[3] which occurs as pollen and ovules develop in the reproductive organs. During meiosis, chromosomal crossover[4] occurs, where parental chromosomes recombine and exchange segments. The resulting chromosomes are then independently sorted into gametes, leading to offspring inheriting different combinations of chromosomes. For example, if there are two pairs of chromosomes, AA′ and BB′, the independent assortment produces four types of gametes—AB, AB′, A′B, and A′B. When crossover is factored in, the variety of gametes increases even further. For rice, which has 12 pairs of chromosomes, the possible combinations of gametes exceed 2^{12} or 4096 types. Moreover, when these gametes randomly combine with counterparts during fertilization, the resulting offspring can have more than 167,772,164,096 (4096 × 4096) unique chromosome combinations. As a result of this process, the chromosomes in the offspring generation are a well-mixed combination of those from the ancestral plant and the cultivated crop. Figure 10.2 simplifies and explains this process.

By analyzing the offspring obtained through this process, researchers identify chromosomal regions containing variations that are present in individuals with crop traits but absent in those with ancestral plant traits. These variable regions are referred to as QTLs. Once QTLs are identified, the next step is to determine which specific genes within these regions are responsible for the observed changes in crop traits. While it is theoretically possible to investigate every gene within the QTLs—potentially hundreds—it is impractical due to the immense time and effort required. Instead, researchers examine prior studies to identify genes already implicated in influencing the crop trait of interest or genes with similar functions. If a gene that appears relevant to the trait is located within the QTL, it is designated as a candidate gene. To confirm its role, the expression of the candidate gene is artificially modified in the ancestral plant or in the crop. If artificially altering the expression of the candidate gene results in corresponding changes in the crop trait, it can be concluded that the genetic variation in that gene determines the trait.

---

[3] A type of cell division that produces sperm and egg cells for sexual reproduction. It cuts the number of chromosomes in half, so each gamete carries only one set, compared to the two sets in the parent cell.

[4] A process during meiosis in which segments of DNA are exchanged between paired chromosomes at specific locations. This shuffling of genetic material creates new combinations of genes, increasing genetic variation in offspring.

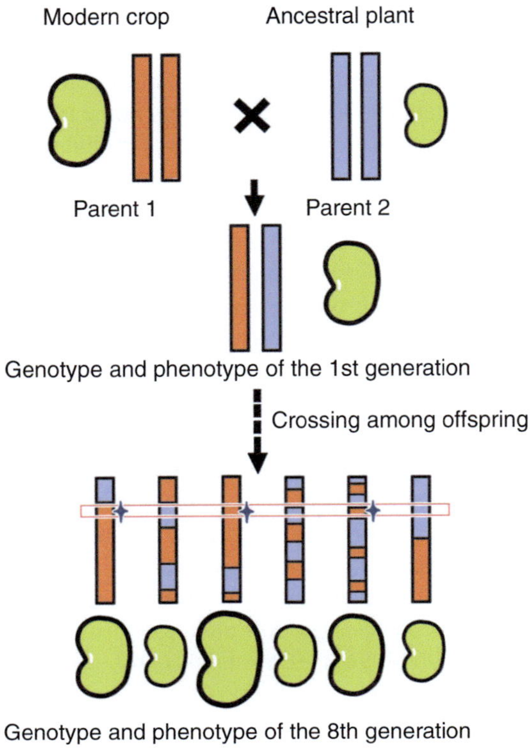

Fig. 10.2 Simplified explanation of quantitative trait locus (QTL) mapping for a domestication gene. This method identifies chromosomal regions associated with traits present in a crop (Parent 1) but absent in its ancestral plant (Parent 2). The long orange box on the genotype of the 8th generation indicates a chromosomal region likely containing the gene responsible for the large size of beans. This conclusion is based on the observation that all large beans in the 8th generation share this chromosomal region (marked as +), which is also present in Parent 1, known for producing larger beans than Parent 2. This region is known as a QTL. To simplify the diagram, only one of the two chromosomes is shown above each bean in the last row

When it is challenging to create hybrids between ancestral plants and crops, GWAS is often employed. This method analyzes the association between specific traits and genetic variations across diverse populations, such as different ecotypes or cultivars, to identify the genes responsible for those traits. The approach involves pinpointing chromosomal regions with variations common in crop individuals but absent in ancestral plant individuals. Among the genes in these regions, candidate genes are selected for further detailed genetic and molecular biological experiments to determine their role in the traits.

GWAS has become an increasingly common method for identifying genes associated with specific traits. Unlike QTL mapping, which requires generating multiple generations of offspring, GWAS analyzes existing cultivars, eliminating the need to create new plants and thus reducing the time required. Previously, GWAS was expensive and required long-term experiments due to the need for sequencing genomes of many ecotypes and cultivars. However, advancements in big data creation and analysis technologies have made GWAS more accessible and efficient. Through such genetic studies, researchers can pinpoint the genetic changes that enable crops to exhibit specific traits.

Examples of Genes Altered During the Domestication Process

By 2020, over 100 genetic changes contributing to crop domestication had been identified through various research methods, including those mentioned above. These changes often involved alterations in the nucleotide sequences of specific genes known as domestication genes. In some cases, larger-scale genetic changes occurred, such as chromosome doubling or quadrupling, or the insertion of transposons.[5]

The functions of proteins encoded by domestication genes are highly diverse. Notably, a considerable proportion of these genes encode transcription factors, which control the expression of other genes. This prevalence can be explained by the fact that losing or drastically altering genes encoding essential proteins can cause plants to die or become severely weakened. In contrast, mutations in transcription factor genes that merely adjust the levels of gene expression are less likely to lead to extreme changes that compromise the plant's survival.

One of the transcription factor genes involved in domestication is the *TB1* (*Teosinte Branched1*) gene in maize, which suppresses the growth of lateral branches. Modern maize has a single, thick stalk, while its ancestor, teosinte, had a highly branched structure (see Fig. 9.1). Interestingly, the *TB1* genes in maize and teosinte encode proteins that have identical amino acid sequences. The difference lies in the increased expression of the *TB1* gene in maize, which results from the insertion of a transposon ("jumping gene") at the *TB1* locus [125]. The increased production of TB1 protein inhibits lateral branch development by promoting the production of plant hormones that suppress branching. Additionally, the protein reduces sugar levels in buds that could otherwise develop into branches, further preventing their growth in maize [31].

Other well-known examples of transcription factor genes altered during domestication are *Sh4* and *qSH1*. These genes determine the degree of seed shattering—the natural dispersal of mature seeds from the parent plant [67]. In wild rice, seed shattering helps spread seeds widely, but this makes it difficult for farmers to collect them efficiently. By contrast, when seeds remain attached to the plant until harvest, they are much easier to gather. In domesticated rice, a single amino acid change in the Sh4 protein led to reduced expression of the *Sh4* gene. Interestingly, the expression of *Sh4* is not completely eliminated but only reduced. This adaptation allows rice seeds to stay attached to the plant during growth but still detach easily from the straw during threshing.

[5] A segment of DNA within the genome that can move from one location to another, often referred to as a "jumping gene."

Fig. 10.3 The abscission layer in rice seeds, the site of seed shattering. The red box highlights the abscission layer, where the *qSH1* gene is expressed and regulates its formation. In japonica rice, a single nucleotide change in the regulatory region upstream of the *qSH1* gene prevents its expression in the abscission layer

To understand the discovery of seed shattering genes, let us examine the research on *qSH1*. Scientists began by conducting experiments to identify QTLs, as described at the beginning of this chapter, to uncover additional genes involved in seed shattering. They used hybrids created by crossing indica rice, which easily sheds its seeds, with japonica rice, which rarely does. After identifying the QTL sites, they analyzed the genes in these regions one by one and identified *qSH1* as being responsible for seed shattering (Fig. 10.3). The *qSH1* gene regulates the formation of the abscission layer[6] at the site where seeds detach. In japonica rice, a single nucleotide change in the regulatory region upstream of the *qSH1* gene prevents its expression in the abscission zone, leading to minimal seed shattering. However, when the *qSH1* gene from indica rice is introduced into japonica rice and expressed, the seeds regain their ability to shatter.

Some crops have twice, four times, or even six times as many chromosomes as their ancestral plants. These crops are known as polyploid crops, with wheat being a prominent example. Normally, in sexual reproduction, offspring inherit half the chromosomes from each parent. However, wheat retains the complete set of chromosomes from both parent plants, resulting in a new species that exhibits the advantages of both. Polyploidy often leads to larger seeds, improved disease resistance, and better adaptability to diverse environments, making polyploid crops more suitable for domestication. Approximately 19% of all crop species have undergone chromosome doubling during their domestication process [84].

[6] A layer of cells formed when leaves, petals, or fruits detach from the plant body.

Harnessing Knowledge of Genetic Changes During Domestication to Breed Better Crops

Research to identify domestication genes is a challenging task that requires time and effort. However, once this knowledge is obtained, it can greatly benefit crop breeding. In some cases, breeding plants to select offspring with desired traits involves growing the plants over a long period. For instance, improving the quality of a fruit-bearing tree usually requires waiting several years for the fruit to form and mature before identifying the individuals with the preferred traits. If the QTLs for those desirable traits are known, plants with genetic variations in those QTLs can be identified at a young stage and selected for further investigation and breeding. This approach reduces the time and effort required to develop a cultivar harboring a trait of interest and is referred to as *marker-assisted breeding*. If the specific domestication genes responsible for the traits are identified, this knowledge can have even broader applications across various fields of biotechnology. The identified genes can be introduced into other crops to confer beneficial traits, or they can be used to investigate whether similar genes in other organisms perform similar functions.

Taking the rice described in the previous section as an example, suppressing the expression of the *qSH1* gene in indica rice could reduce seed shattering, potentially increasing rice yields. To support such efforts, genotype–phenotype databases have been developed, containing information about genes that confer useful traits for crops. For instance, scientists successfully identified a QTL associated with submergence tolerance in rice [139], pinpointed the causative gene in that region [140], and used marker-assisted breeding to enhance submergence tolerance in rice.

The Impact of Agriculture on Humans 11

Relationships formed in social life influence both parties involved. While humans have altered plants through agriculture, agriculture has also brought tremendous changes to human society. The large population of humans (8 billion and counting!) currently inhabiting the Earth owes much to domesticated crops. Through the cultivation of crops, humans gained a stable food supply, which enabled massive population growth [28, 45]. The increase in population and the complexity of society have led to remarkable advancements in culture and knowledge.

However, the increase in the human population has also led to major challenges, including global warming and environmental destruction. To support the growing population, humans continuously improved crop productivity, resulting in a tremendous increase in food production. This, in turn, fueled further population growth and the demand for essential resources such as food, clothing, housing, and goods. Most of these resources are derived from plants, and their excessive use has depleted Earth's resources to the point where natural recovery has become increasingly difficult. While this book does not explore these issues in depth, it is important to note that the development of agriculture has been a major contributor to the global warming crisis of the twenty-first century.

Even more regrettable is that, despite the continued increase in agricultural production, many people around the world still suffer hunger. During the Green Revolution of the twentieth century, scientists succeeded in doubling or tripling the productivity of wheat and rice. However, at the same time, the global population grew by three to four times over the past century, leaving many people on Earth still suffering from a lack of food.

Y. Lee, B. Y. Choi, *Social Lives of Plants*,
https://doi.org/10.1007/978-981-96-9837-0_11

Eliminating hunger, ensuring that the majority of people can enjoy a desired standard of living, and stabilizing the Earth's environment are challenges humanity must address in the twenty-first century. To effectively tackle these challenges, research in multiple areas is essential. New crops and agricultural methods must be developed, and ways to minimize humanity's burden on nature while aiding its recovery need to be identified. In the realms of politics, society, and economics, strategies must be devised to create stable and happy societies without further increasing the human population.

12 Human-Induced Changes in the Global Environment and Plants

Global Warming and Its Causes

Currently, global warming and climate change are occurring at an unprecedented speed. Plenty of scientific evidence supports that it has been caused by "greenhouse gases" emitted through human activities [73]. Natural factors that could strongly influence changes in the Earth's surface temperature, such as solar activity, have remained relatively stable over the past 9000 years. Similarly, volcanic activity within the Earth has shown little variation over the past 2500 years. In contrast, the concentrations of major greenhouse gases produced by human activities—carbon dioxide, methane, and nitrous oxide—have increased by 47%, 156%, and 23%, respectively, compared to pre-industrial levels in the 1750s, reaching the highest levels in Earth's history over the past 800,000 years. The greenhouse gases in the atmosphere have led to the greenhouse effect, resulting in a rise in the Earth's surface temperature. The average global surface temperature between 2001 and 2020 was 0.99 °C higher than that of the early industrial era in 1850.

The rise in Earth's surface temperature has led to serious changes in various global environmental and climatic conditions. For instance, the ice sheets in the Antarctic and Arctic regions are melting, and as a result, sea levels in 2018 were 0.2 m higher compared to 1901. Consequently, low-lying island nations in the Pacific and Caribbean regions have experienced coastal flooding, leading to a reduction in their territory. It is predicted that in the future, their populations may have to relocate elsewhere. In addition, since the 1950s, the frequency of extreme summer heatwaves, heavy rainfall, and typhoons has increased, while the occurrence of severe cold spells and heavy snowfall during winter has become less common.

What human activities cause greenhouse gas emissions? Energy consumption for industries, transportation, and heating or cooling buildings accounts for 70% of all greenhouse gas emissions. Surprisingly, agriculture, livestock farming, food production, and distribution are the second-largest source, contributing 25% of total emissions [107]. Converting grasslands or forests into farmland releases carbon

Y. Lee, B. Y. Choi, *Social Lives of Plants*,
https://doi.org/10.1007/978-981-96-9837-0_12

stored in the soil and destroys forest resources that naturally absorb carbon dioxide. This happens by removing the plants that store carbon and then by disturbing the soil, which speeds up the breakdown of organic matter and releases carbon dioxide. Even rice cultivation, a staple in many Asian countries, contributes to greenhouse gas emissions. Rice paddies require large amounts of energy for irrigation. Moreover, irrigated rice paddies are sources of methane, a potent greenhouse gas. In irrigated soils, oxygen concentration is low, and when there is not much oxygen, the usual microbes that break down organic matter cannot do their job. Instead, other microbes take over and slowly decompose the material in a way that produces methane instead of carbon dioxide. Furthermore, the production of nitrogen fertilizers, which are widely used to enhance crop growth, also contributes to greenhouse gas emissions. The production process of nitrogen fertilizers alone accounts for 1%–2% of total greenhouse gas emissions.

In short, human activities—especially agriculture and industry—to support the rapidly growing population have led to massive greenhouse gas emissions. These emissions accumulate in the atmosphere year after year, further accelerating global warming.

Future Impacts of Rising Atmospheric Carbon Dioxide on Plant Growth

From 10,000 years ago until the eighteenth century, before the Industrial Revolution, the atmospheric carbon dioxide concentration remained around 280 ppm. However, by May 2022, it had risen to 421 ppm. If greenhouse gas emissions continue at the current rate, it is projected that by the year 2100, the atmospheric carbon dioxide concentration will exceed 1000 ppm. How will this condition affect plant growth?

If only the carbon dioxide concentration increases while other conditions remain the same, plant growth will be stimulated. Carbon dioxide is a raw material for photosynthesis. For plants to grow, they must convert carbon dioxide into sugars, which are then used to build cell walls and other structural materials. However, often, even when light is available, photosynthesis cannot occur due to insufficient carbon dioxide levels inside the leaves. Plants obtain carbon dioxide through small openings called stomata on the surface of their leaves (Fig. 12.1). When plants open their stomata, they can absorb carbon dioxide, but at the same time, they lose water. This occurs because the humidity in the atmosphere is almost always lower than inside the plant, causing water to diffuse from the plant to the atmosphere. Water loss poses a risk to plants; excessive water loss can cause leaves to wilt, and prolonged wilting may lead to plant death. As atmospheric carbon dioxide concentrations rise, the risk of water loss in plants decreases, since sufficient CO_2 can be absorbed with minimal stomatal opening. This physiological response reduces transpiration and is expected to enhance plant growth. Accordingly, plant photosynthetic productivity is projected to increase by 2100 compared to pre-industrial levels, while the exact magnitude of this increase remains uncertain.

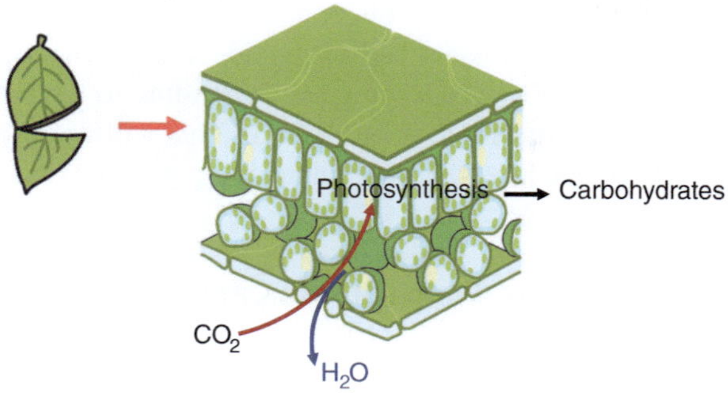

Fig. 12.1 A schematic diagram of a stoma and its role in gas exchange in a leaf. The lower surface of a leaf has small openings called stomata, which allow air to pass in and out. When stomata open, the plant absorbs carbon dioxide (CO_2) but also loses water (H_2O) in the process. Since excessive water loss can be harmful, plants regulate their stomatal apertures tightly. As a result, the carbon dioxide concentration inside the leaf is often too low for optimal photosynthesis

However, rising atmospheric carbon dioxide levels will also bring about several other changes that negatively affect plant growth. One of the most important factors is temperature. If greenhouse gas emissions continue at the current rate, raising atmospheric carbon dioxide concentration above 1000 ppm by 2100, the Earth's surface temperature is projected to rise by approximately 4 °C [24]. Many plants including the world's largest crops are highly sensitive to high temperatures, and such plants may not grow well even if carbon dioxide levels increase.

Another challenge is plant nutrition. In addition to carbon dioxide, plants require various nutrients for growth, with nitrogen and phosphorus, in particular, being essential. However, the availability of these nutrients on the Earth's surface is insufficient to support unlimited plant growth [136]. During the early stages of global warming, the amount of available nitrogen is expected to increase due to accelerated decomposition of organic matter in the soil caused by rising temperatures, which enhances nitrogen recycling [82]. However, this increase will be short-lived, and over the long term, nitrogen sources in the soil are expected to become depleted [36], leading to a gradual decline in plant growth. The case of phosphorus presents an even greater limitation. It takes several decades to centuries for phosphorus in the soil to transform into forms that plants can absorb. This slow transformation process is expected to limit phosphorus supply, potentially suppressing the rapid plant growth driven by rising carbon dioxide levels. Consequently, phosphorus limitation is anticipated to become an even more critical constraint than nitrogen deficiency on plant growth.

Therefore, despite a large increase in atmospheric carbon dioxide partial pressure by 2100, photosynthetic productivity is not expected to rise markedly. As a result, the amount of carbon dioxide fixed by plants will not increase significantly. On the other hand, as global warming raises surface temperatures, the respiration rates of living organisms will rise significantly, and the decomposition of organic matter in the soil will accelerate. Consequently, the carbon dioxide once stored in the soil will be released into the atmosphere at a faster rate. In other words, while global warming is unlikely to enhance carbon dioxide fixation through photosynthesis, it will promote respiration and decomposition, leading to the rapid conversion of stored carbon into atmospheric carbon dioxide. This will cause a sharp rise in atmospheric carbon dioxide levels, further intensifying global warming.

Future Impacts of Rising Atmospheric Carbon Dioxide on Agriculture

If rising carbon dioxide levels do not enhance photosynthesis in forest trees, how will they affect agricultural productivity? Scholars predict that increasing carbon dioxide and the resulting climate changes will reduce agricultural yields. Why? Higher carbon dioxide levels lead to rising global temperatures, increasing the frequency of droughts and floods, which hinder crop growth and reduce agricultural productivity.

It is expected that developing countries will experience a more severe decline in crop yields compared to developed countries. This is because rising temperatures and changing climates will alter the water, fertilizer, and environmental conditions required for crop growth, and developing countries, with limited access to technology and capital, are likely to struggle to adapt to these changes in time. For instance, as temperatures rise, more soil moisture will evaporate into the atmosphere drying the soil, which, in turn, will necessitate additional irrigation. However, installing irrigation systems requires a large financial investment. In addition, climate changes may necessitate the cultivation of new crop varieties suited to the altered conditions, and new weeds and pathogens may pose a threat to crops. Addressing these challenges will be difficult for developing countries. In summary, global warming is expected to bring major changes to agriculture, with poorer nations facing the greatest difficulties and the most severe impacts.

The rise in atmospheric carbon dioxide will bring qualitative changes to agricultural products. Crops will contain more sugars but lower amounts of proteins and essential fatty acids.[1] This will make foods taste sweeter but less nutritious, potentially leading to a situation where people feel full without receiving balanced nutrients. This phenomenon is referred to as "hidden hunger." The reason for the decline in proteins and other nutrients relative to sugars is that under higher carbon dioxide concentrations, plants do not need to produce as much protein to perform photosynthesis as they did before. Currently, under low carbon dioxide conditions, plants produce large amounts of RuBisCO, a key protein that captures carbon dioxide for photosynthesis. In spinach leaves, for example, RuBisCO accounts for 40% of the total protein content. However, under higher carbon dioxide levels, such large amounts of this enzyme are no longer necessary. Other proteins required for

[1] A type of fatty acid required for normal growth, development, and health in humans and many animals, but which cannot be synthesized by the body. These must be obtained through the diet and include omega-3 and omega-6 fatty acids.

photosynthesis are also needed in smaller quantities. As a result, the overall protein content in plants is expected to decrease.

On the other hand, as global warming leads to harsher growing conditions for plants, they may produce more secondary metabolites [12]. These chemical compounds help plants defend themselves, and some have been used by humans as medicines. When environmental conditions deteriorate and stress levels rise, the production of flavonoids in vegetables and phenolic compounds in quinoa increases [30]. Additionally, higher carbon dioxide concentrations have been found to enhance the production of morphine in poppies, which is used as a pain reliever, and digoxin in *Digitalis*, which is used as a heart medication [148].

Future Impacts of Rising Atmospheric Carbon Dioxide on Plant Diversity

Plants growing in natural environments are also vulnerable to rapid environmental changes because they cannot relocate their habitats. Recently, due to the rapid changes in temperature and other environmental conditions, plant species that can adapt well to environmental stress are surviving, while those that cannot are becoming extinct, leading to a decline in plant species diversity [108]. Additionally, plant interactions with insects play an important role in changing species diversity. Insects like bees and moths transfer pollen, and many plants depend on them for reproduction. Approximately 80% of the 350,000 species of flowering plants worldwide depend on insects for pollination and reproduction [93]. However, due to global warming, the timing of plant flowering and the emergence of insects may become misaligned, making it difficult for plants to receive pollen and reproduce.

Research has shown that the increased frequency of extreme heat events due to global warming has led to a sharp decline in honeybee populations [123]. Experts in South Korea have established that unusual weather patterns played a considerable role. For instance, in September and October 2021, low temperatures hindered the development of worker bees, while unusually high temperatures in November and December caused flowers to bloom in the southern regions, allowing overwintering bees to get out. However, due to subsequent cold temperatures and exhaustion, many bees failed to return to their hives and perished [20]. Honeybees are not only essential pollinators for many wild plants but also for major agricultural crops. Therefore, if honeybees were to disappear due to global warming, many crops and wild plant species would face the risk of extinction.

Plants that rely exclusively on a single insect species for pollination will fail to produce seeds if that insect does not emerge from its eggs at the right time. In this changing environment, plants that utilize a variety of insects for pollination will have a reproductive advantage, while those that do not depend on insects for reproduction will be even more advantaged. For instance, plants such as pine trees, which produce vast amounts of pollen and rely on wind for pollination instead of insects, will have a competitive edge.

In short, atmospheric carbon dioxide levels are rising too rapidly. As a result, plants that cannot tolerate the new climate conditions—or the changes these conditions cause in the organisms they depend on—are at risk of extinction. This will reduce plant diversity.

The Rate of Plant Extinction

Over the past 100 years, the global human population has grown three to four times. To secure food for people and livestock, forests have been cleared for farmland, and to make space for housing, natural environments have been heavily altered. As a result, many plant and animal habitats have been lost, leading to frequent extinctions. According to a report published by the Royal Botanic Gardens, Kew, in June 2019, a total of 571 plant species have disappeared from the Earth over the past 250 years. This number is more than twice the combined total of extinct birds, mammals, and amphibians, which stands at 217 species. Currently, the rate of plant extinction is approximately 500 times faster than the natural rate of extinction without human influence.

For example, the Saint Helena olive (*Nesiota elliptica*), which once thrived on the South Atlantic island of Saint Helena, experienced habitat loss due to deforestation and grazing by introduced goats, leading to a severe population decline. It was believed to be extinct by the nineteenth century; however, in 1977, a single remaining tree was discovered on the island. Despite efforts by botanists to preserve it, the tree died in 1994, and the young trees propagated from its cuttings also perished in 2003, resulting in the species' complete extinction. The genetic material of this plant is now preserved in the DNA bank at the Royal Botanic Gardens, Kew. Another example is the Chile sandalwood (*Santalum fernandezianum*), a tree known for its fragrant wood. It grew on the Juan Fernández Islands off the coast of Chile and was heavily logged for its aromatic timber, starting in 1624. By the late nineteenth century, only a few individuals remained. The last known Chile sandalwood tree was documented in a photograph taken in 1908, and no specimens have been found since.

In South Korea, several plant species are at risk of extinction, including Korean lady's slipper (*Cypripedium japonicum*), which grows in Gangwon and Jeolla provinces; *Aconitum austrokoreense* Koidz., which is found at elevations below 600 m in Gyeongbuk, Gyeongnam, and Jeonnam provinces; and Korean fir (*Abies koreana*), which inhabits high mountain areas such as Mt. Halla and Mt. Jiri [87]. Without strong conservation efforts, plant extinction rates will continue to rise. If these plants provide food or habitat for animals, those species will also face a high risk of extinction.

Efforts to Preserve Agricultural Biodiversity

As the global environment changes daily, it is uncertain which crops will reliably provide us with food in the future. Thus, preserving the genetic diversity of crops is essential. Seed banks and seed vaults serve this purpose by serving as a repository

Fig. 12.2 Exterior view of the Svalbard Global Seed Vault (Photo by Crop Trust, via Flickr. Licensed under CC BY-SA 2.0)

of the seeds of existing crops to maintain their genetic diversity. One of the best-known facilities is the Svalbard Global Seed Vault (Fig. 12.2), operated by the Norwegian government. This seed vault was created to safely store seeds and preserve them for future generations, even in the face of catastrophic events like nuclear war, natural disasters, or rapid climate change that could cause mass plant extinctions.

The Svalbard Global Seed Vault is located deep inside a mountain on Spitsbergen island, about halfway between mainland Norway and the North Pole. The facility extends more than 100 m into the mountain and sits 130 m above sea level to protect it from flooding or other potential disasters. To ensure the long-term preservation of seeds, the vault maintains both low temperature and low humidity. It is built into permafrost, and electric cooling systems maintain it at −18 °C. The seeds are dried before storage and placed in a sealed, dry environment. Because of the region's consistently cold climate, the frozen seeds are expected to remain protected even in the event of a power outage. However, despite being located in a geologically optimal area for seed preservation, rapid global warming has posed challenges. In 2017, melting permafrost in the Arctic, caused by accelerated climate change, led to flooding at the entrance of the vault. As the pace of environmental change exceeds expectations, the future of seed vaults has become uncertain.

The Svalbard Global Seed Vault was inaugurated in 2008 and is equipped with facilities capable of storing a total of 4.5 million seed varieties, with 500 seeds per variety, for a total capacity of 2.5 billion seeds. To date, the vault has received and stored over one million seed samples from various countries. South Korea has also deposited a total of 23,185 seed samples from 44 different crops in the Svalbard Global Seed Vault [62]. Only the countries or research institutions that originally deposited the seeds have the authority to retrieve them. The first instance of seeds being withdrawn from the vault occurred due to the Syrian civil war. In 2015, scientists from the seed bank of the International Center for Agricultural Research in the Dry Areas, based in Aleppo, Syria, requested access to drought- and heat-resistant wheat seeds that had been deposited in the vault since 2012. Due to the ongoing war, they were unable to continue their research in Syria. After retrieving the seeds, they

Fig. 12.3 The Baekdudaegan Global Seed Vault in Bonghwa, Gyeongbuk, South Korea

relocated their research efforts to Lebanon and Morocco, where they resumed their studies [132]. Many crops grown in these regions have strong resistance to drought and high temperatures. Research into the genetic factors responsible for this stress tolerance could help develop crops that ensure stable food supplies amid global warming.

South Korea also has its own seed vault, the Baekdudaegan Global Seed Vault (Fig. 12.3). Opened in 2017 in Bonghwa, Gyeongbuk Province, the seed vault is currently managed by the Baekdudaegan National Arboretum. It is the only seed vault in the world that preserves not only crop seeds but also wild plant seeds. The seeds stored in this facility have been collected from various institutions, including domestic arboreta and universities. As of March 2021, a total of 95,395 seed samples from 4751 plant species have been stored, including 2166 native plant species, 329 rare plant species, and 180 endemic plant species. The facility is classified as a national security site, and public access is restricted.

Efforts to Preserve Overall Plant Diversity

Life on Earth has been evolving for 4 billion years, with organisms forming complex relationships and coevolving to create intricately connected ecosystems. For an ecosystem to remain stable, its living organisms must carry out essential life processes that help stabilize Earth's water, air, and soil. From a human perspective, the biological resources within ecosystems are like treasures in a vault—once lost, they cannot be restored. The environmental changes we face today are unfolding at an unprecedented pace and scale, far beyond anything our ancestors experienced. As a result, it is difficult to predict the future based on historical knowledge. In such uncertainty, we need to safeguard this treasure trove of biodiversity. Even plants that may seem of little value today could become valuable resources in the future. Therefore, it is necessary to ensure their preservation and prevent their extinction.

To preserve plant diversity, it is necessary to minimize the conversion of forests into farmland and to leave natural ecosystems as intact as possible. To achieve this, agricultural methods and crops must be improved to increase productivity per unit area, ensuring that sufficient food can be produced on limited farmland. Among various technologies, smart farming has emerged as a potential solution. For example, growing wheat in a 10-story smart farm[2] could achieve hundreds of times the productivity compared to a traditional farm of the same area [4]. However, smart farms face economic challenges due to high energy consumption for artificial lighting and the substantial costs of facility maintenance and labor.

Another proposed way to reduce farmland use and restore the environment is adopting a vegetarian diet, which encourages reducing consumption of animal-based foods like meat, milk, and eggs. Producing beef, for instance, generates 31 times more carbon dioxide than producing tofu from soybeans. While other animal-based foods like pork, chicken, milk, and cheese produce less carbon dioxide than beef, their emissions are still much higher than those of plant-based foods like rice, potatoes, wheat, and soybeans. The reason why animal-based food production generates more carbon dioxide is due to the substantial energy loss that occurs throughout the process of feeding livestock with pasture and grains, which are then consumed by humans. For example, only 5% of the calories from livestock feed are converted into beef, while the remaining 95% is lost. To produce 1 kg of beef, approximately 20 kg of feed, such as corn and soybeans, is required.

According to a 2018 report, 80% of the world's farmland is used to grow pasture and feed crops for livestock [102]. If the entire human population adopted a completely plant-based diet, consuming no meat or animal-derived products, it is estimated that only one-quarter of the farmland currently in use would be sufficient to feed the global population. Even if people stopped consuming only beef and lamb, farmland use could be reduced by half. Reducing the amount of farmland used for livestock would allow the reclaimed land to be returned to wild plants and animals. However, meat consumption is currently on the rise. If this trend is not reversed in favor of more plant-based diets, restoring the global environment will be challenging. If the global population were much smaller, the planet might sustain high levels of animal product consumption. However, with eight billion people depending on current agricultural practices to produce animal-based foods, the Earth's land, freshwater, ocean, and atmosphere will inevitably continue to deteriorate [106].

Can Meat Alternatives Help Preserve Plant Diversity?

While adopting a vegetarian diet is an effective way to protect the global environment, people still have a strong desire to eat meat. Moreover, the growing emphasis on protein intake for health makes it challenging to popularize vegetarianism. To overcome this challenge, meat alternatives have emerged as a way to provide

[2] A farm that uses information and communication technology (ICT) to monitor, analyze, and control the plant growth environment. Based on real-time data, smart farms automatically adjust conditions such as temperature, light, and humidity to optimize plant growth, improving efficiency and productivity.

protein without relying on animal products. The meat alternative industry is growing as a venture sector in the United States, Europe, and Japan, and South Korea has recently identified meat alternative as one of its promising industries.

There are three types of meat alternatives: plant-based meat substitutes, cultured meat, and edible insects. Plant-based meat substitute is made by using proteins derived from plants such as soy and wheat to replicate the appearance and taste of animal-based foods. For example, since 2020, the global fast-food franchise McDonald's has collaborated with the company Beyond Meat to introduce the McPlant, a plant-based meat alternative burger patty made from a blend of peas, rice, potatoes, and beets, replacing traditional beef patties. However, plant-based meat alternatives still differ greatly in taste and texture from real meat, which has hindered their widespread adoption. As a result, McPlant is not yet widely available at all McDonald's locations.

Recently, cultured meat technology has been developed, which involves growing animal cells such as muscle and fat cells to produce meat. Proponents of cultured meat claim that it can significantly reduce greenhouse gases like carbon dioxide and methane compared to traditional livestock farming. While it is still uncertain to what extent cultured meat technology can be commercialized, numerous startups worldwide are conducting research with the goal of making it commercially viable. To achieve commercialization, cultured meat must be produced in a way that closely mimics the taste of conventional meat while also reducing production costs to make it affordable. The most expensive part of production is the culture medium, which must contain amino acids, essential fatty acids, sugars, and growth factors necessary for cell growth. These nutrients are typically produced using microbes such as *E. coli* or yeast, which themselves require a growth medium—ultimately derived from crops. This means that even if meat can be produced in bioreactors, the nutrients and energy required for its production will still originate from crops. Consequently, reducing the cost of cultured meat to below that of conventional meat remains challenging. Furthermore, the cultured meat industry is unlikely to drastically reduce global farmland use, as it will still depend on crops, either directly or indirectly. Therefore, even if cultured meat becomes commercially available, its contributions to environmental restoration and plant diversity conservation are expected to be limited. However, since cultured meat can be produced without slaughtering animals, it may contribute to improving animal welfare by reducing the number of animals raised and slaughtered for food.

Insects are highly efficient at accumulating protein in their bodies, making them a promising protein source. Producing protein from insects requires only about one-tenth of the land and feed needed for traditional livestock farming. Therefore, if edible insects become widely accepted, they could help preserve plant biodiversity by reducing the demand for agricultural land. In some parts of the world, insects are already part of the diet. For example, in Korea, during past periods of food scarcity, people collected and consumed grasshoppers from rice fields. Even today, silkworm pupae are commonly sold as food in amusement parks and traditional markets.

However, insects remain an exotic food ingredient rather than a primary protein source. Many people still have an aversion to eating them, and like plant-based meat alternatives, edible insects differ greatly in taste and texture from conventional meat. As a result, they have yet to gain widespread acceptance as a viable alternative protein source.

The Future of Plant-Human Interactions 13

Can We Grow Plants in Space?

As concerns over extreme climate change, severe environmental pollution, and the risk of unexpected disasters grow, people have begun to dream of leaving Earth to live on other celestial bodies. With recent advancements in rocket technology, the prospect of humans reaching and settling on nearby celestial bodies, such as the Moon and Mars, is becoming increasingly realistic. But is it truly possible to terraform the Moon and Mars into habitable environments where humans can reside for extended periods? For terraforming to succeed, it would be essential to cultivate food crops and establish a stable food supply, just as on Earth. However, the space environment beyond Earth is vastly different from our planet's conditions, making crop cultivation a major challenge. So, what factors must we consider to successfully grow plants in space?

For living organisms to survive, a suitable temperature must be maintained. On the Moon, the difference between the highest and lowest daily average temperatures is about 273 °C, while on Mars, the variation is about 93 °C. Plants cannot survive such extreme temperature fluctuations, so in order to cultivate plants on the Moon and Mars, greenhouses capable of maintaining stable temperatures would be required. Greenhouses would also serve the important role of protecting plants from high-energy cosmic radiation.[1] Additionally, a large amount of water is required for plant cultivation. Surprisingly, recent explorations suggest that both the Moon and Mars contain substantial water reserves, though in forms that are difficult to use. If this water can be efficiently extracted and processed, it may be possible to secure the water needed for growing plants.

In addition to water, plants need light and carbon dioxide for photosynthesis. Both the Moon and Mars receive sufficient sunlight to support plant cultivation;

[1] High-energy particles and electromagnetic radiation originating from outer space, traveling at speeds close to the speed of light.

Y. Lee, B. Y. Choi, *Social Lives of Plants*,
https://doi.org/10.1007/978-981-96-9837-0_13

however, the availability of carbon dioxide differs between them. While the atmospheres of both the Moon and Mars are extremely thin, there is a key difference in their carbon dioxide availability. Mars has a thin atmosphere composed of about 95% carbon dioxide, which is sufficient for plant growth. In contrast, the Moon's atmosphere is so tenuous that carbon dioxide exists only in trace amounts, so carbon dioxide would have to be brought in from outside to grow plants.

Gravity is another important factor for plant growth. Plants sense the direction of gravity and extend their roots downward into the soil to absorb the inorganic nutrients necessary for growth. However, the gravitational pull on the Moon and Mars is weaker than on Earth. While Earth's gravity is defined as 1G, Mars has a gravity of 0.38G, and the Moon's gravity is even lower at 0.17G. Can plants develop their roots properly under such conditions? Laboratory experiments have artificially simulated the gravitational conditions of Mars and the Moon to study root growth. The results indicated that under Martian gravity, root development was similar to that observed under Earth's gravity. However, in the case of plants germinated under lunar gravity, root cell division increased, but overall cell growth slowed. These findings suggest that for plants to develop roots in a manner similar to that on Earth, a gravitational force stronger than that of the Moon is required [81].

How about soil? Do Mars and the Moon have soil suitable for plant growth? Both Mars and the Moon have sand-like soil, which contains a considerable amount of nutrients necessary for plant growth. However, this soil has almost no nitrogen sources and contains high levels of aluminum and chromium, which are harmful to plant growth. A study by Dr. Wamelink showed that plants could germinate and survive for 50 days without fertilizer in soil that mimicked the surface layer composition of Mars and the Moon [134]. Nonetheless, some crops did not germinate at all, and only 2 out of 14 plant species bloomed and bore fruit. Another experiment tested the actual soil of the Moon for its ability to support plant growth. A research team led by Professor Ferl at the University of Florida conducted a study using *Arabidopsis* plants grown in lunar soil brought back by the Apollo mission. When *Arabidopsis* seeds were sown in lunar soil and supplied with nutrient solution daily, they germinated but did not grow well and showed extreme stress responses [98].

As demonstrated by these experimental results, protecting plants from the toxic components of lunar and Martian soil and supplying necessary fertilizers appear to be the primary challenges. In particular, the problem of how to provide nitrogen sources, which are essential for protein synthesis, is a difficult one. In the movie *The Martian*, which depicts human survival on Mars, there is a scene where the protagonist resolves the nitrogen deficiency issue by using crew members' excrement as fertilizer for plants. Since the crew members consumed food brought from Earth, the nitrogen ultimately came from crops originally harvested on Earth and was being recycled. Due to the unsuitability of lunar and Martian soil for plant growth, some argue that Earth soil must be transported in order to cultivate crops. However, carrying soil all the way to the Moon and Mars seems highly impractical. The recent discoveries confirming the presence of a substantial amount of water on the Moon and Mars raise the possibility of hydroponic cultivation using this water. However, unlike Earth's rivers or oceans, the water on the Moon and Mars does not exist in a readily accessible liquid form. On the Moon, water is found as ice in permanently shadowed regions near the poles, where sunlight never reaches. On Mars, water also exists mainly as ice at the poles and beneath the surface. Thus, extracting a sufficient amount of water presents a technological challenge. Even if water could be extracted for hydroponics, the necessary fertilizers would still need to be produced on Earth and transported.

Finally, the absence of symbiotic partners for plants is also an important issue. On Earth, plants do not exist in isolation; they coexist with other living organisms, including neighboring fungi, plants, insects, and herbivores, forming complex ecological relationships. This raises the question of how plants could survive on the Moon or Mars, where such interactions are absent. In an experiment where the sweet clover (*Melilotus officinalis*) was grown in soil that mimicked the surface layer of Mars, researchers found that the plants grew better when nitrogen-fixing bacteria were introduced [46]. This suggests that symbiotic partners may be necessary for optimal plant growth on other planets.

Overcoming these numerous challenges and achieving self-sufficiency in agricultural production on other planets will require an unprecedented level of technological advancement. No matter how much technology progresses, plants on other planets will still need to grow within human-made greenhouses. However, since recreating an environment identical to Earth's is highly challenging, plants will need to be extensively modified to grow and produce in the new space environment.

The more research is conducted on growing plants in extraterrestrial environments, the more it becomes clear that there are numerous and extremely difficult challenges to overcome. At the same time, it highlights just how precious Earth's environment is. This raises the question: rather than focusing on planning human life on other planets and cultivating plants in space, should not we prioritize protecting and restoring Earth's environment?

What Kinds of Plants Will We Cultivate in the Future?

The number of crops cultivated by humans so far is very small compared to the total number of plant species growing on our planet. Currently, approximately 150 species are actively cultivated, while around 7000 species are consumed to some extent though they have not been domesticated to crops. Given that there are about 400,000 known plant species, the proportion of plants utilized by humans is quite limited. Among the 150 cultivated crops, humans obtain most of their calories from a small number of species. Corn, rice, and wheat collectively account for approximately 50% of the global caloric intake. In total, 15 crop species—including corn, rice, wheat, potatoes, soybeans, sugar beets, sugarcane, bananas, cassava, sweet potatoes, tomatoes, onions, barley, apples, and watermelon—contribute approximately 70% of the calories consumed worldwide [35]. The fact that humans utilize only a small fraction of plant species compared to the total number available suggests that there is potential to develop many other plants into new crops.

Scientists around the world are working to develop new crops for various reasons. First, there is a need for crops that can be cultivated with less reliance on fertilizers and pesticides, which heavily burden the environment. Second, it is necessary to develop crops that use water efficiently. Agriculture consumes a large amount of water, and water scarcity is a serious global issue, as droughts often lead to decreased agricultural production. Third, there is a need for crops capable of providing a stable food supply under rapidly changing climate conditions such as rising global temperatures, increased frequency of droughts and floods, and higher ozone levels in the atmosphere. The impact of climate change on agriculture is already evident. In South Korea, for example, areas suitable for apple production have gradually shifted northward because apples with high sugar content—preferred by consumers—require warm days and cool nights. Conversely, in the southern coastal regions, subtropical crops such as pineapples and apple mangoes can now be cultivated. As climate change continues, efforts to develop new crops will be crucial. Fourth, from a global perspective, the human population continues to grow, and as living standards rise, people are consuming more meat. To meet this demand, more crops must be produced to serve as animal feed. New crops need to be developed not only for food and feed but also for other uses. As the population grows, more raw materials from crops will be required to produce various goods. For example, there will be an increasing demand for crops that serve as raw materials for industrial products such as textiles, paper, and biomaterials.

Plants that have not yet been domesticated as crops likely possess traits that made them unsuitable for cultivation. However, by applying new technologies—particularly molecular biology techniques—these traits can be modified, allowing these plants to be developed into useful crops for food, industrial applications, or even for mitigating global warming.

Development of New Food Crops

Koreans consume a much wider variety of plants, including a broader range of wild vegetables and seaweeds, than those commonly consumed in other countries. Recently, the government has provided research funding to support efforts to develop better cultivation methods for wild mountain vegetables. These diverse plant species have the potential to supply various nutrients and medicinal compounds.

Seaweed is highly nutritious and has the potential to become a widely consumed food worldwide. Currently, only a small fraction of the global population eats seaweed. Many Westerners, for example, tend to dislike seaweed because of its slimy texture. A mathematician I met in the United States once said, "Korean cucumber chilled soup tastes good, but I just couldn't bring myself to eat miyeok-guk (seaweed soup) because of its slimy texture." It is said that food preferences are largely established in childhood and are quite difficult to change in adulthood. However, as global cultural exchanges become more frequent and the numerous benefits of seaweed gain recognition, its consumption is likely to increase.

There are also efforts to cultivate microalgae for use in sustainable food, pharmaceuticals, and industrial materials. Single-celled algae consist of 40%–60% protein by weight and contain all the amino acids essential for human health. They can also produce vitamins and other functional compounds. Already, microalgae species such as *Chlorella*, *Euglena*, and *Spirulina* have been introduced as health supplements in Japan and Korea. Currently, venture companies in Israel, Canada, the Netherlands, and New Zealand are developing technologies to cultivate microalgae for protein production and are working toward commercializing microalgae-derived protein [16]. Microalgae protein has the potential to replace dairy-derived protein and can also be blended into various protein-based products. While microalgae have long been considered an alternative food and fuel source, high cultivation costs have limited commercialization. However, recent advancements in cultivation technologies, including improved lighting methods, have made it possible to produce microalgae protein at a price comparable to that of protein derived from soybeans or peas.

Crop species resistant to environmental stresses such as high temperatures, droughts, floods, and high salinity are expected to be used more frequently in the future, playing a greater role in sustaining the global food supply. For example, quinoa can grow in saline soils, making it a promising crop for broader cultivation. Additionally, as interest in and demand for meat alternatives continue to rise, there is a growing possibility that not only crops for livestock feed but also crops specifically cultivated for producing cultured meat ingredients will be widely grown.

Development of Crops to Mitigate Climate Change

In 2017, Dr. Joanne Chory at the Salk Institute in the United States launched the *Harnessing Plants Initiative* (HPI), an ambitious project aimed at developing crops that can help mitigate global warming. Following Dr. Chory's passing in 2024, the HPI team has been led by her colleague, Dr. Wolfgang Busch. Their research focuses on creating crops capable of storing and sequestering carbon dioxide—a major contributor to global warming—in the soil. Plants already play a significant role in removing vast amounts of carbon dioxide from the atmosphere through photosynthesis. Globally, plants absorb 746 billion tons of carbon dioxide annually. However, when plants die and decompose, 727 billion tons of carbon dioxide are released back into the atmosphere. Additionally, human activities contribute 37 billion tons of carbon dioxide emissions each year. When these factors are combined, the total amount of atmospheric carbon dioxide increases by 18 billion tons annually.

The HPI team works to change three important plant traits to help fight global warming. Their first target has been to create plants that make more suberin[2] in their roots. Suberin is a complex polymer that demands a lot of resources from the plant

[2] A water-repelling, complex biopolymer made mainly of long-chain fatty acids and polymerized glycerol. It is the main component of cork, such as that used in bottle stoppers. In plants, suberin forms a protective barrier around root vascular tissues, blocking the uncontrolled movement of water and solutes through spaces between cells. Instead, essential substances are selectively trans-

for its production. These resources come from plant photosynthesis, which uses carbon dioxide as the raw material. Therefore, plants that produce more suberin could help lower the levels of carbon dioxide in the atmosphere. Additionally, suberin is highly resistant to decomposition. If plants produce more suberin, the carbon stored in their tissues will take longer to be released back into the atmosphere after they die. Their second target is to develop plants with larger root systems, while the third is to create plants with roots that grow deeper. Plants with both larger and deeper roots would absorb and store more carbon dioxide from the atmosphere. This stored carbon would then be buried deeper in the soil, reducing the chance of it quickly returning to the air. According to their research, even a 2% increase in the total amount of carbon dioxide fixed by plants could offset most of the carbon dioxide emissions from human activities and slow down the pace of climate change.

The HPI team conducts experiments with *Arabidopsis thaliana*, a model plant that is easy to genetically modify, to achieve these goals. Once they have successful results with *Arabidopsis*, they plan to use the same techniques on major crops like soybeans, wheat, corn, and rice. You might wonder why they focused on crops to combat global warming. To effectively reduce climate change by capturing carbon dioxide with plants, it is essential to cover a large area with vegetation. However, given the world's growing population and limited land, dedicating vast areas solely for this purpose is difficult. Crops, on the other hand, are already extensively cultivated worldwide. If farmers plant crop varieties designed to help tackle climate change, they will not only produce food but also help lower the levels of carbon dioxide in the atmosphere. Dr. Chory expressed optimism in her TED Talk, saying, "Plants are going to do it for us, all we have to do is give them a little help, and they will go and get a gold medal for humanity."

The goal of the HPI team is admirable if they can achieve it, but ensuring that crops develop larger root systems and produce more suberin while maintaining high agricultural yields will be extremely challenging. In the mid-twentieth century, scientists achieved "the Green Revolution" by developing high-yielding crops, saving many people from starvation. However, those crops required large amounts of fertilizers and water to increase productivity, which not only boosted agricultural output but also contributed to rising atmospheric carbon dioxide levels. HPI team aims to engineer crops that work harder—reducing carbon dioxide concentrations in the air while maintaining agricultural productivity—an extremely challenging task. The team acknowledges these challenges but believes that advances in plant genetics have made their goals achievable. As evidence, they already identified genes that promote larger root growth and successfully engineered *Arabidopsis* plants with large root systems. However, beyond the scientific hurdles, there is another major challenge: convincing farmers to adopt these newly developed crop varieties in their fields. Despite these challenges, the fact that scientists are dedicating their efforts to

---

ported via specialized transporter proteins in the cell membrane. This barrier function allows plants to absorb beneficial nutrients while blocking harmful substances.

such ambitious projects offers a beacon of hope for the future of our planet, and one cannot help but hope that this type of research continues actively.

Recently, seaweed has been proposed as a potential crop to help mitigate climate change (Fig. 13.1). Using seaweed to capture carbon dioxide from the atmosphere is significantly more cost-effective than chemical methods, since photosynthetic organisms utilize solar energy to absorb carbon dioxide. Although trees also absorb carbon dioxide, land suitable for tree growth is often used for other purposes, making it expensive, and trees take a long time to grow. In contrast, some species of seaweed can grow up to 1 m per day. Unlike terrestrial plants, seaweed does not have stems or roots and performs photosynthesis with its entire body. It requires no fertilizers, pesticides, or irrigation. Since 70% of Earth's surface is covered by oceans, theoretically, planting seaweed in just 9% of the ocean's area—equivalent to about 4.5 times the size of Australia—could offset all human carbon dioxide emissions. If harvested seaweed, which has absorbed carbon dioxide as part of its biomass, is stored in deep ocean waters over 1 km below the surface, where oxygen levels are too low for organic decomposition, the captured carbon dioxide could remain trapped in the ocean for thousands of years without returning to the atmosphere. In addition to carbon dioxide absorption, seaweed can also capture nitrogen

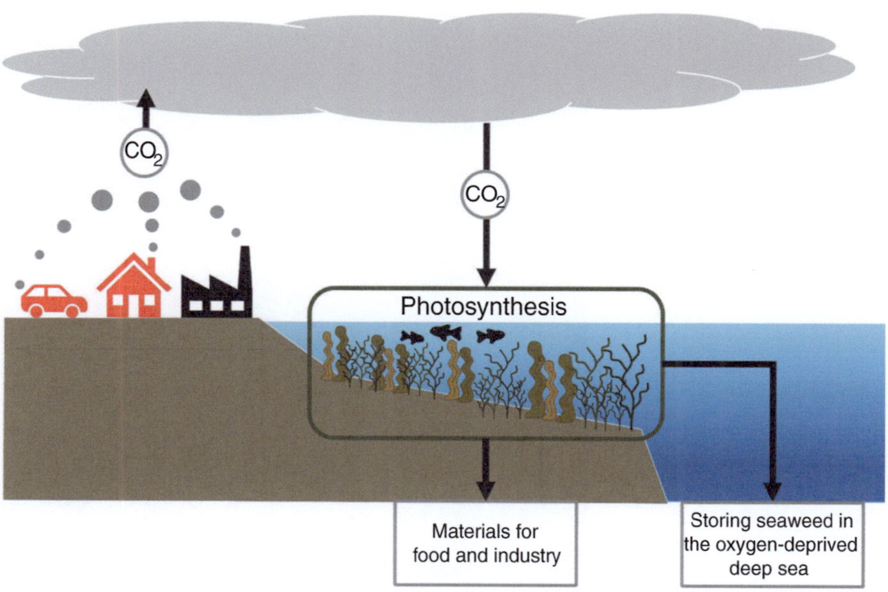

Fig. 13.1 The idea of cultivating seaweed to absorb carbon dioxide. Growing seaweed on the coastal shelf allows it to absorb carbon dioxide from the atmosphere and grow rapidly. When the cultivated seaweed is stored in the deep sea, where oxygen is absent, organic matter does not decompose, preventing the release of carbon dioxide back into the atmosphere. This process effectively removes carbon dioxide from the atmosphere and stores it stably in the deep sea. Additionally, areas where seaweed grows can support the growth of fish and shellfish, providing a valuable source of protein. Furthermore, seaweed can be utilized for industrial purposes. (This figure illustrates the concept proposed by Dr. Flannery)

from agricultural runoff. By cultivating seaweed along coastal regions, it is possible to create environments that support fish and shellfish, fostering sustainable and eco-friendly fisheries that provide high-quality protein for both humans and animals. Furthermore, seaweed can serve as a raw material for bioplastics[3] and biofuels,[4] contributing to the development of sustainable industries.

Development of Industrial and Medicinal Crops

Some crops are used as industrial raw materials, and the variety of such crops is likely to expand in the future. For example, *Jatropha curcas*, a tropical plant (Fig. 13.2), has traditionally been used as a natural fence to keep livestock contained because its unpleasant odor deters animals from approaching. However, it was recently discovered that oil extracted from its seeds can be used as fuel, making it an increasingly popular biofuel crop.

Jatropha can grow well in poor, nonarable land without requiring large amounts of water or fertilizer, and it is highly drought resistant. In Africa, there have been efforts to cultivate *Jatropha* as a source of energy, with the hope that selling its oil could help alleviate poverty. However, since this tree has not been widely cultivated

Fig. 13.2 *Jatropha*, a promising crop for biofuel production (Photo by francisco3_, via iNaturalist. Licensed under CC BY 4.0)

[3] A type of plastic derived from renewable biological sources such as plants and microorganisms. Unlike petroleum-based plastics, some bioplastics break down more easily, which may lessen environmental impact.

[4] A type of fuel derived directly or indirectly from biological sources such as plants, animals, microorganisms, algae, and organic waste. During combustion, biofuels release carbon dioxide that was originally absorbed from the atmosphere by plants through photosynthesis. Because of this recycling of carbon, biofuels are considered a form of renewable energy.

as a crop, farmers lack knowledge about the best cultivation methods to achieve consistently high yields and how to properly manage the plant. If future research can address these challenges, *Jatropha* could become a valuable source of income for people in Africa.

Camelina sativa is another oil-producing plant that is highly resistant to environmental stress. It belongs to the Brassicaceae family, which includes crops like cabbage and radish. *Camelina* seeds contain approximately 40% oil by weight. *Camelina* oil is not only a healthy cooking oil but also serves as a raw material for various industrial applications.

There have been attempts to extract lipids from microalgae for energy production. However, the processes of cultivating microalgae, extracting oil, and converting it into fuel are costly. As a result, biofuels derived from microalgae remain more expensive to produce than petroleum or coal, hindering their commercialization to date.

Rather than using plants for fuel production, there is growing interest in utilizing them to produce high-value proteins. The production of high-value pharmaceuticals in plants is known as *molecular farming* or *biopharming*. For example, researchers are developing antigen proteins for vaccines, growth hormones, and insulin in plants, with several successful cases already reported. Notably, a COVID-19 vaccine was developed using spike proteins of the coronavirus produced in plants and was approved for human use. Medicago, a Canadian biopharmaceutical company, collaborated with GlaxoSmithKline, a British pharmaceutical company, to develop Covifenz, a COVID-19 vaccine in which spike proteins were expressed in plants and then isolated and purified. The vaccine successfully completed phase three clinical trials involving 24,000 adults in six countries and was approved for use in Canada in February 2022.

There are also efforts to produce bioplastics from plants. Currently, most plastics are derived from petroleum, but starch and lipid compounds obtained from plants can also be used to produce plastic. Plastics made from such plant-based materials are called bioplastics. Bioplastics have a lower carbon footprint[5] than petroleum-based plastics because their raw materials are synthesized through photosynthesis. As of 2016, bioplastics accounted for only 1% of all plastics, but with advancements in technology, their market is expected to grow. However, the extensive use of starch and lipids from food crops for consumer goods could worsen food shortages. To mitigate this issue, dedicated bioindustrial crops should be developed. Similar to *Jatropha*, these new crop varieties should be capable of growing in harsh conditions without relying on arable farmland, ensuring a sustainable bioindustry that does not compromise the human food supply.

[5] The total amount of greenhouse gases released into the atmosphere—directly or indirectly—by a person, organization, activity, or product over a specific period.

Crops Engineered with Foreign Genes

Since the late twentieth century, genetic engineering has advanced rapidly, enabling scientists to introduce and express foreign genes in plants (Fig. 13.3). Crops developed through this process are called genetically modified organisms (GMO) or living modified organisms (LMO). The term LMO emphasizes that the organism is capable of reproduction, whereas GMO includes both reproductive and nonreproductive organisms. Currently, the most widely used genetically modified crops are those that have been engineered with a pest-resistant gene derived from the bacterium *Bacillus thuringiensis* and crops that have been modified to resist herbicides.

Pest-Resistant Crops

The soil microorganism *Bacillus thuringiensis* is a pathogen of silkworms. Farmers have traditionally used this bacterium to reduce damage from silkworms by spraying it onto crops. *Bacillus thuringiensis* produces Cry proteins, which, when ingested by pests, break down in their highly alkaline digestive tracts and create pores in the gut lining. These pores allow gut bacteria to spread throughout the insect's body, triggering a strong immune response that ultimately kills the insect. Instead of applying *Bacillus thuringiensis* itself, scientists have introduced the *Cry* gene directly into plants through genetic engineering, creating crops specifically designed for pest resistance. These genetically modified crops are known as Bt crops. Currently, Bt cotton, Bt potatoes, and Bt corn are widely cultivated worldwide. Because these crops can defend themselves against pests, farmers can reduce pesticide use while the crops maintain high yields without suffering from pests.

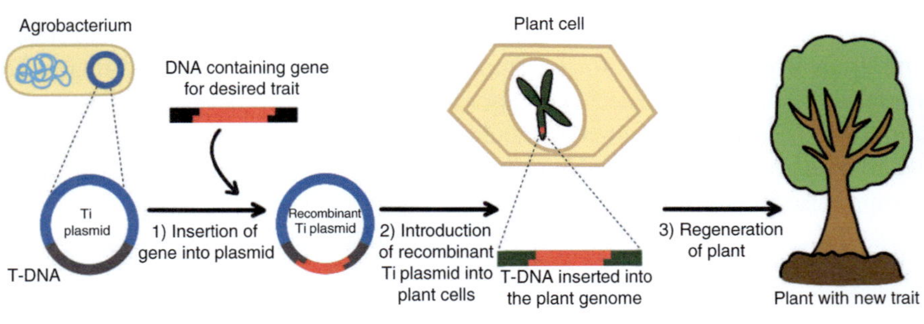

Fig. 13.3 Method for creating GMOs using *Agrobacterium*. Scientists insert a desired gene into the T-DNA, a small genomic fragment used by *Agrobacterium*, and then introduce this modified T-DNA into the plant. The T-DNA is stably integrated into the plant's genome, allowing the foreign gene to be expressed in the mutant plant, which then exhibits the desired trait

Herbicide-Resistant Crops

One of the most important tasks in farming is weed control, as weeds compete with crops and reduce yields. Traditionally, farmers plow their fields before sowing seeds to remove weeds. In Korea, women farmers often wear towels on their heads and use homi (a traditional Korean hand plow) to weed their fields. If weeds continue to grow excessively, the last resort is to apply herbicides. Among the many herbicides used in gardens and farmland, glyphosate, the active ingredient in the herbicide Roundup, is one of the most widely used. Glyphosate inhibits the activity of an enzyme essential for the synthesis of three amino acids: phenylalanine, tyrosine, and tryptophan. When Roundup is applied, plants cannot produce these amino acids, which ultimately causes them to die. This herbicide has minimal impact on animals because they do not produce phenylalanine and tryptophan but obtain them from their diet. Additionally, animals generate tyrosine through a different biosynthetic pathway that is not affected by glyphosate.

Herbicide-resistant GMO crops have been genetically engineered to express a bacterial gene that produces an enzyme not affected by glyphosate. This modification allows the crops to synthesize the three essential amino acids even in the presence of glyphosate. When farmers grow glyphosate-tolerant GMO crops and apply glyphosate-based herbicides, the herbicide selectively kills weeds while leaving the crops unharmed. This technology reduces the need for plowing and manual weeding, helping to lower labor costs and decrease soil erosion associated with conventional tillage. However, due to patent protections, farmers are legally prohibited from saving and replanting GMO seeds. As a result, they must purchase new seeds each season—a requirement that has sparked debate over whether the primary beneficiaries of these technologies are multinational corporations. In response to this criticism, companies often state that they have implemented management practices to prevent the spread of herbicide-resistant genes into natural ecosystems and to prevent the development of herbicide-resistant weeds. In the United States, a large proportion of soybeans and corn cultivated today are herbicide-resistant GMOs, and they are widely used in animal feed and processed foods.

Nutrient-Enriched GMO Crops

Crops engineered for pest and herbicide resistance have been beneficial to farmers, but for consumers, their primary advantage has been lower food prices, offering little direct benefit beyond that. This is one reason why many consumers remain skeptical about GMOs. To address this limitation, genetic engineering is now being used to develop nutrient-enriched crops that offer direct health benefits to consumers. Notable examples include vitamin A-enriched rice and flavor-enhanced tomatoes. Vitamin A-enriched rice, developed by Dr. Potrykus' research team at ETH Zurich, aims to help people suffering from vitamin A deficiency, which can lead to weakened immunity and vision problems. While vitamin A deficiency is not a major concern in developed countries, where people consume meat, leafy vegetables, and tropical fruits rich in vitamin A, it remains a serious problem in developing countries. Alarmingly, about 30% of preschool-aged children worldwide are estimated to be vitamin A deficient [137]. To address this, the research team has focused on increasing the beta-carotene content of rice, a staple food in many developing countries. Since beta-carotene, a precursor to vitamin A, is converted into vitamin A when consumed, regularly replacing ordinary rice with beta-carotene-enriched rice could potentially eliminate vitamin A deficiency. The researchers introduced two genes into rice: one encoding a phytoene synthase enzyme from corn and another encoding a phytoene desaturase enzyme from bacteria. These genetic modifications allowed rice grains to accumulate yellowish beta-carotene pigment, producing a new cultivar known as Golden Rice [96]. Golden Rice contains over 30 μg/g of beta-carotene, and studies have confirmed that its consumption significantly alleviates vitamin A deficiency [25].

However, despite its potential benefits, Golden Rice has faced strong opposition from anti-GMO activists for more than 20 years, which has prevented its widespread adoption. Opponents argued, without scientific basis, that Golden Rice could threaten biodiversity, increase reliance on GMOs, and pose higher health risks than traditional rice varieties. To dispel these concerns, multiple research institutions, including the US FDA, conducted safety assessments. Studies analyzing the DNA composition of Golden Rice and allergic responses after consumption demonstrated that Golden Rice is no more harmful to human health than conventional rice [92]. After more than 20 years of regulatory hurdles, Golden Rice was approved for consumption in Canada and the United States in 2018, and in 2021, the Philippines became the first country to approve its commercial cultivation and distribution. Following the success of vitamin A-enriched Golden Rice, research is now underway to increase the vitamin B content of cassava and rice to help populations suffering from vitamin B deficiencies [74, 103].

However, many GMO crops have yet to be commercialized, primarily due to controversies and opposition stemming from concerns that GMOs may be harmful to the environment and human health. As a result, proving their safety requires enormous financial investment. According to Bayer, a multinational corporation currently developing and selling GMO crops, the development of a single GMO crop takes over 16 years and costs around 115 million USD. A large portion of these astronomical costs and lengthy development timelines is spent on navigating regulatory hurdles associated with GMOs. Due to these challenges, only a handful of large multinational corporations have been able to bring GMOs to market. Meanwhile, small research institutions and universities, despite successfully developing beneficial GMO crops, often struggle to move beyond the research phase and achieve commercialization.

Excessive Concerns About the Environmental Impact of GMOs

Are GMO crops truly as dangerous as some people fear? First introduced in the early 1990s, GMO crops have been cultivated for over 30 years. Many initially worried that GMOs might negatively impact human and animal health or harm the environment, but no such cases have been reported. On the contrary, studies have shown that the use of pest-resistant GMO crops has led to a reduction in overall pesticide use, contributing to positive environmental effects [86]. Some environmental groups have expressed concerns that genes for pest or herbicide resistance in GMO crops could spread to wild plants, potentially creating "superweeds" that are impossible to eliminate and could overrun ecosystems. However, such an event has not yet occurred. One reason this is highly unlikely is that GMO crops are not well adapted to surviving in the wild without human intervention. Because GMO plants constantly expend resources to express the genetically introduced traits, they have less resources available for growth and survival in competitive natural environments. In nature, resources are always limited, and when plants allocate energy to one function, they must reduce investment in others. For example, GMO crops that continuously produce pest-resistant proteins grow slower when no pests are present, making them less competitive in the wild. As previously discussed, most plants in nature have evolved to prioritize growth in favorable conditions and activate defense mechanisms only when a threat arises. This strategy enhances their ability to compete with other plants in natural ecosystems. Therefore, the idea that GMO crops could outcompete other plants in the wild and disrupt ecosystems is highly improbable.

Some have argued that GMOs are unsafe simply because they are artificially created by scientists. However, the mechanism used to produce GMOs is actually a natural process that occurs in the wild environment. One common method for creating GMO plants involves using *Agrobacterium tumefaciens*. This bacterium naturally transfers its genes into plant cells, causing crown gall disease. *Agrobacterium* carries a small piece of DNA known as T-DNA, which it transfers to the plant during infection. Once inside a plant cell, the T-DNA is inserted into the plant's genome, where it becomes a permanent part of the plant's genetic material and begins to function within the plant. If scientists want to introduce a specific gene into a plant, they can insert that gene into the T-DNA region of *Agrobacterium*. By infecting the plant with the modified *Agrobacterium*, the target gene becomes part of the plant's genome and can be stably expressed (Fig. 13.3). Amazingly, this same process—which scientists use to create GMOs—occurred naturally thousands of years ago in the ancestors of modern sweet potatoes. Researchers have discovered that the genome of today's sweet potato contains T-DNA sequences derived from *Agrobacterium*, and the genes from these T-DNA fragments are still expressed in sweet potatoes today [71]. This means that the sweet potato we have been eating for at least 1000 years is a natural GMO. The fact that humans have consumed sweet potatoes for millennia without issues suggests that using T-DNA to create GMOs is not inherently harmful.

In 2016, 107 Nobel laureates issued a statement urging Greenpeace and other environmental organizations to end their unfounded opposition to GMOs. They declared that there is no scientific evidence to show that GMO crops are more harmful to human or animal health than conventionally bred crops. Instead, they pointed out that GMO crops actually have a reduced environmental impact compared to traditional crops and contribute positively to global biodiversity. Given these insights, claims that currently cultivated GMOs harm human health or the environment are not supported by scientific evidence.

Gene-Editing Technology and Its Future Potential

CRISPR-Cas9 gene-editing technology is currently the most efficient way to precisely change genes in living organisms. This tool has revolutionized biotechnology. In recognition of their work in elucidating the mechanism of CRISPR-Cas9 and developing genome-editing techniques based on it, Professors Jennifer Doudna of the University of California and Emmanuelle Charpentier of the Max Planck Institute in Germany were awarded the 2020 Nobel Prize in Chemistry.

The CRISPR-Cas9 system is a defense tool that microorganisms use to guard against viruses. When a virus infects a microorganism, the microorganism captures and stores fragments of the viral DNA within a specific region of its genome called CRISPR (clustered regularly interspaced short palindromic repeats). If the same virus invades the microorganism again, the stored DNA sequence in CRISPR is used to produce a complementary RNA[6] molecule. This RNA guides the Cas9 enzyme (CRISPR-associated protein 9, a nuclease) to the invading viral DNA, allowing Cas9 to precisely cut and degrade the viral genome. Through this process, the microorganism effectively defends itself against viral infections.

As its name suggests, the CRISPR-Cas9 gene-editing technology utilizes the Cas9 nuclease, which cuts DNA, and a guide RNA (gRNA) that directs the enzyme to its target site. The Cas9 nuclease protein forms a complex with the gRNA, which leads the Cas9 nuclease to a complementary DNA sequence. Once the gRNA aligns with its DNA target, Cas9 cleaves the double-stranded DNA at the designated site (Fig. 13.4). When DNA is cut, scientists can use the cell's natural DNA repair mechanisms to modify the nucleotide sequence at the cleavage site. Because CRISPR-Cas9 gene-editing technology recognizes DNA sequences through gRNA, it allows precise targeted cutting at specific locations within an organism's entire genome, regardless of its complexity.

If scientists want to modify a different gene, they only need to change the sequence of the gRNA. Recent advancements in chemical synthesis technologies have made it simple and efficient to synthesize custom RNA sequences. Many biotechnology companies now offer RNA synthesis services at relatively low costs, allowing researchers to easily modify gRNA sequences and edit various target genes with ease. Due to these advantages, CRISPR-Cas9 technology has rapidly spread worldwide and is now widely used in biotechnology laboratories. Today, many biological research labs utilize CRISPR-Cas9 gene-editing to precisely edit genomes in various organisms and conduct a wide range of genetic studies. In particular, CRISPR-Cas9-based genome editing has been actively researched in the plant

[6] An abbreviation for ribonucleic acid, a macromolecule essential for carrying, regulating, and expressing genetic information. RNA is made using DNA as a template, so its sequence is complementary to the DNA strand.

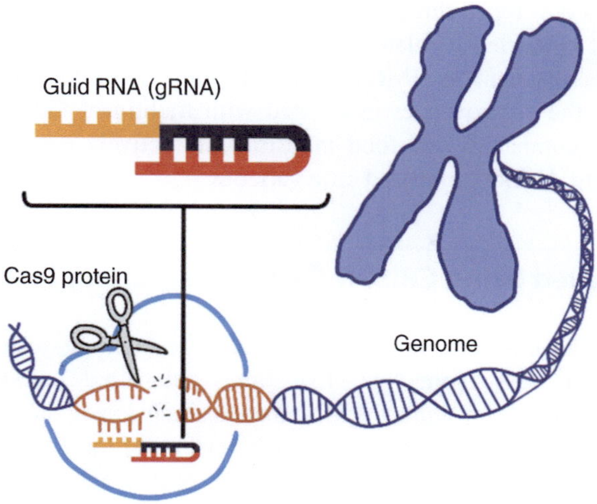

Fig. 13.4 Genome editing using CRISPR-Cas9 gene-editing technology. The gRNA binds to the Cas9 nuclease and directs it to a DNA sequence complementary to the gRNA. The Cas9 nuclease then cuts the double-stranded DNA like scissors. After the DNA is cleaved, the cell's natural DNA repair mechanisms kick in to fix the break. Scientists introduce insertions or deletions during the repair process, thereby creating genetic modifications

sciences. This is because, unlike traditional *Agrobacterium*-mediated genetic modification, CRISPR-Cas9 does not introduce foreign genes into plants. Instead, it precisely cuts and modifies specific nucleotide sequences within the plant's existing genome, making targeted genetic alterations.

The regulatory status of CRISPR-Cas9 gene-edited plants that do not contain foreign genes varies by country. While some nations classify these plants as GMOs, others do not. In the United States, the world's leading agricultural producer and a major GMO crop cultivator, regulatory agencies prioritize assessing the safety of the final product rather than how the mutation was created. The US Department of Agriculture (USDA) has already decided that plants developed using CRISPR-Cas9 gene-editing technology are not subject to GMO regulations. Similarly, Australia and New Zealand have announced that plants with simple gene deletions through the CRISPR-Cas9 system will not be considered GMOs. In Japan, crops modified using CRISPR-Cas9 gene-editing techniques are also not classified as GMOs. However, in Europe, where GMO regulations are stricter, oversight is based on how the mutation was introduced. Under EU regulations, plants harboring any genome modification that could not have occurred naturally are classified as GMOs. As a result, gene-edited plants using the CRISPR-Cas9 system are subject to the same regulatory framework as traditional GMOs in Europe [72]. In South Korea, clear regulatory guidelines for gene-edited crops using the CRISPR-Cas9 system have not yet been established. As of now, these crops are regulated under the same framework as GMOs. However, discussions are underway to revise the law to exempt CRISPR-Cas9-derived plants from additional risk assessments if they are deemed as safe as crops developed through traditional breeding or natural mutations [60]. Since different countries regulate CRISPR-Cas9-derived crops differently—and some do not classify them as GMOs—plants developed using this technology may avoid some of the controversies associated with traditional GMOs. As a result, many breeding companies and food industries are actively exploring the use of CRISPR-Cas9 to develop improved crop varieties.

Crops Improved Using CRISPR-Cas9 Gene-Editing Technology

The first crop developed using CRISPR-Cas9 gene-editing technology was a tomato enriched with more than five times the usual amount of GABA (γ-aminobutyric acid), an amino acid known to help lower blood pressure and reduce stress. According to early tasters and reviewers, the flavor of the tomato remained largely unchanged. This outcome was expected, as GABA itself has little to no taste, and the gene editing targeted only the GABA transaminase enzyme to increase GABA levels—without altering any genes related to flavor. This tomato was developed by Sanatech Seed, a venture company originating from the University of Tsukuba in Japan, and has been commercially available in Japan since September 2021.

There has been a proposal to use gene-editing technology for *de novo* domestication, the process of creating new crops from the wild ancestors of established crops [141]. Among the wild ancestors of modern crops, some species have been overlooked due to their low yield or poor taste. However, many of these wild plants possess resistance to environmental stresses such as high temperatures, droughts, and pests, which are becoming more frequent due to climate change. The idea behind *de novo* domestication is to introduce mutations into genes that control yield and nutritional value in these wild ancestors to develop superior crop varieties. Many genes that changed during the domestication of modern crops have already been identified, and genome-editing techniques have been advanced. Therefore, conceptually, editing the genomes of wild ancestors to create new crops is not complicated. While the domestication of many of today's crops took thousands of years, using CRISPR-Cas9 gene-editing technology for *de novo* domestication could produce new crops in just a few decades. This approach has already been successfully used to develop improved wild tomatoes and wild rice varieties as crops [142, 149].

Is Vertical Farming a Viable Alternative for the Future of Agriculture?

In 1999, Professor Despommier of Columbia University proposed the concept of vertical farming, an idea aimed at reducing the amount of land needed for agriculture by growing crops in multiple stacked layers inside buildings. To cultivate plants in indoor multitiered systems, shelves are installed to hold the crops, and lighting fixtures are placed above each shelf to provide artificial light (Fig. 13.5). Additionally,

Fig. 13.5 Vertical farming facilities require complex equipment and highly trained personnel. Vertical farming uses specialized systems—such as artificial lighting, climate controls, and nutrient dispensers—to grow crops indoors. For vertical farming to become commercially viable on a larger scale, the cost of artificial lighting must be reduced, and technologies must be developed to automate system management, thereby reducing high labor costs

the system requires temperature and humidity control, along with water and nutrient delivery methods, to sustain plant growth.

The concept of vertical farming is simple, but operating an actual vertical farm requires precise technology and skilled personnel. If there is even a brief power outage or if a pump fails disrupting the water supply, plants may wither or their growth could be altered, leading to substantial losses. Therefore, technicians must always be on-site to ensure that all mechanical systems function properly, and agricultural experts are needed to monitor and optimize plant growth (Fig. 13.5). As a result, this farming method incurs high labor costs. Proponents of vertical farming argue that stacking crops in multiple layers can reduce the need for farmland, allowing some areas to be returned to nature, benefiting both humans and the environment. Additionally, growing crops in urban areas eliminates the need for long-distance transportation of agricultural products, reducing energy consumption and carbon footprint. Other advantages of vertical farming include minimizing pest and disease risks, reducing water usage, and enabling year-round production of fresh food, regardless of weather conditions.

However, vertical farming is currently limited to the cultivation of leafy vegetables and has not yet produced high-calorie crops on a large scale. What challenges must be overcome for this method to enable sustainable large-scale crop production while minimizing carbon emissions? Aside from social issues, such as conflicts with traditional farmers, that may arise, an even more important issue here is the economic viability of the technology. First, advanced automation technologies are needed to reduce labor costs, particularly those associated with agricultural experts. Second, it is essential to develop cost-effective and environmentally friendly artificial lighting. The economic advantage of traditional crop cultivation lies in the fact that plants utilize free solar energy. When artificial lighting replaces sunlight, energy consumption increases, potentially leading to higher carbon dioxide emissions than conventional farming methods. Currently, leaf vegetables like lettuce are the primary crops grown in vertical farms because they require less light than other crops, making artificial lighting costs more manageable. To expand vertical farming, affordable, efficient, and sustainable lighting solutions must be developed. If an artificial sun could be realized, providing unlimited energy and light, this issue could be resolved. Another key challenge is recreating the ecological interactions that plants experience in natural environments. In traditional agriculture, plants interact with other plants, microorganisms, insects, and animals, which influence their growth, flavor, and nutritional composition. Such biotic interactions stimulate plants to produce secondary products, critical for food quality and nutrition. Therefore, for vertical farming to succeed, solutions must be developed to artificially simulate these natural interactions within a controlled environment.

In South Korea, vertical farming has recently gained attention, driven by the expectation that this method could become a key pillar of future agriculture and precision farming. Since South Korea is a global leader in LED technology and possesses advanced IoT (Internet of Things) capabilities, it is well-positioned to take the lead in vertical farming innovations by integrating these technologies. One example is LG's home plant cultivation device, "Tiiun," which can be considered a type of vertical farming appliance for households. Additionally, the Korean startup "N.Thing" has developed container-based vertical farm modules and is exporting them to the Middle East, where consumers are interested in growing fresh vegetables locally. Furthermore, BioApp, a South Korean biotechnology company specializing in plant-based vaccine production, also utilizes vertical farming systems for plant cultivation.

Is Environmentally Friendly Agriculture Possible?

Preserving the environment while ensuring a stable and abundant food supply is a central and formidable challenge of our century. To protect the environment, a diverse range of plant species must be left to grow naturally. However, agriculture requires clearing natural vegetation and planting only a single crop species in a given area. When land is plowed for farming, the microbial ecosystems and small insects that thrive there are disrupted, and the soil becomes more vulnerable to erosion by wind and rain. Farmers use pesticides to protect crops from harmful pathogens and insects, but these chemicals also eliminate beneficial microorganisms and insects. Consequently, completely environmentally friendly agriculture is an almost unattainable ideal. Instead, increasing crop yield per unit area can allow more land to be returned to nature, reducing environmental destruction. From this perspective, GMO crops that exhibit higher yields and resistance to pests and weeds may actually be beneficial for the environment [86].

To preserve the environment, it is necessary to develop farming methods that minimize environmental damage. With the emergence of the concept of "sustainable agriculture," numerous ideas have been introduced. Specific approaches include crop rotation; planting nitrogen-fixing crops such as soybean, clover, or alfalfa during the off-season to enhance soil fertility; and no-till farming, where seeds are planted in small holes without plowing the entire field to prevent soil erosion. Additionally, combining livestock farming with crop production allows animal manure to be used as fertilizer. Planting diverse plant species alongside crops helps support pollinators and natural enemies of pests, as demonstrated by "the push-pull strategy" developed in Africa.

However, these sustainable methods are still mostly theoretical, and many farmers stick to traditional agricultural practices. The main reasons are that switching to sustainable farming is complex, requires a lot of knowledge and money, and does not always lead to better productivity than conventional methods. For sustainable agriculture to be widely adopted, it must ensure stable income for farmers while also producing enough food to feed the global population. However, it remains uncertain whether both goals can be achieved simultaneously. Nonetheless, given the serious impacts of global warming and climate change, agriculture cannot continue to operate in ways that contribute to these crises. Ongoing technological progress and research are essential for making sustainable farming a viable reality. Key steps include developing smart farming techniques for precise monitoring and management of soil nutrients and plant health; leveraging accurate weather forecasts to support farm-level decisions; equipping farmers with reliable agricultural information and effective tools; and cultivating environmentally friendly crops. A revolutionary transformation in agriculture is needed to protect and restore the planet's environment.

Appendix

Recently Developed Techniques in Biochemistry, Genetics, and Phylogenetics

1. *DNA Sequencing Technology*

DNA sequencing, a technique used to decode the genetic information contained in DNA, has advanced dramatically, enabling the acquisition of genetic data from numerous organisms. The first DNA sequencing method, known as Sanger sequencing, was developed in the 1970s by the British biochemist Dr. Frederick Sanger. This technique involves setting experimental conditions that allow DNA polymerase to synthesize DNA while introducing dideoxy nucleotides, which inhibit normal DNA synthesis. The incorporation of these modified nucleotides causes random termination of the polymerization reaction, generating DNA fragments of various lengths. By analyzing these fragments, the nucleotide sequence of DNA can be determined. Since Sanger sequencing can decode only a single DNA sequence per analysis, sequencing entire genomes, which contain vast amounts of genetic information, required substantial time and cost.

Entering the 2000s, DNA sequencing technology advanced to a new level with the development of next-generation sequencing (NGS). This technology miniaturized and automated Sanger sequencing reactions, enabling large-scale parallel processing of sequencing data. When analyzing the same amount of DNA, the cost of NGS is only about one-thousandth of that of traditional Sanger sequencing. This revolutionary advancement reduced sequencing costs while allowing for the analysis of vast amounts of DNA sequences. As a result, researchers have been able to study transcriptomes and genomes that were previously difficult to analyze.

DNA sequencing technology continues to evolve, and third-generation sequencing methods have emerged. Among these, Pacific Biosciences' single-molecule real-time (SMRT) sequencing technology enables the analysis of long DNA sequences in a single run. This technique allows the key process of DNA synthesis to occur within a space of approximately 20 zeptoliters (10^{-21} L), enabling real-time observation of a single DNA polymerase incorporating nucleotides at the molecular level. This advancement has improved the efficiency of DNA sequencing. Whereas

Y. Lee, B. Y. Choi, *Social Lives of Plants*,
https://doi.org/10.1007/978-981-96-9837-0

next-generation sequencing methods could read approximately 300 base pairs (bp) per experiment, SMRT sequencing technology can analyze DNA sequences exceeding 10,000 bp in length.

While most commercially available DNA sequencing methods rely on DNA polymerase-mediated DNA synthesis, another third-generation sequencing technology, nanopore sequencing, analyzes DNA sequences without DNA synthesis by leveraging the physicochemical properties of nucleotides. The four nucleotides that make up DNA—adenine, guanine, cytosine, and thymine—have distinct chemical structures and physicochemical characteristics, which are used to determine DNA sequences. In nanopore sequencing, a flow cell separates an electrolyte solution using a highly resistive membrane, into which a nanopore protein is embedded. This nanopore protein allows ions to pass through, generating a constant electrical current. To sequence DNA, single-stranded DNA molecules are introduced into the flow cell, where they pass through the nanopore protein. Because each nucleotide has unique physicochemical properties, it disrupts the ionic flow differently as it passes through the nanopore protein, producing distinct current signals. By analyzing these current fluctuations, the DNA sequence can be determined. In theory, nanopore sequencing does not limit DNA length if the DNA remains intact. In practical experiments, this method can analyze DNA sequences averaging over 13,000 bp in length. Furthermore, since nanopore sequencing does not require DNA polymerase-mediated replication, it can directly analyze DNA modifications such as DNA methylation. Third-generation sequencing technologies enable the sequencing of genomes from organisms whose DNA is too long to be accurately read using previous sequencing methods.

2. *Metabolomics*

Metabolomics is a research field aimed at identifying the complete set of metabolites present in a living organism. This research includes investigations into the types of chemical compounds found within an organism, their concentrations, biosynthetic pathways, degradation processes, etc. With advancements in analytical techniques and data acquisition technologies, metabolomics has led to the discovery of numerous new metabolites. However, achieving a comprehensive understanding of metabolites in organisms will take considerable time due to the vast number of chemical compounds in living organisms, their complex structures, and the wide range of concentration differences. While it has been estimated that plants contain around one million different metabolites, some researchers suggest that the actual number may be much higher.

Several techniques are used in metabolomics, each with its own advantages and limitations. Among these, ultra-high-performance liquid chromatography-high-resolution mass spectrometry (UHPLC-HRMS) has recently gained considerable attention. UHPLC utilizes a long column packed with tiny particles (less than 2 microns in diameter) to separate chemical compounds as the sample passes through the column. HRMS then applies a high voltage to ionize the separated compounds, and the resulting ions are further separated based on their mass-to-charge ratio, generating a mass spectrum. By analyzing this mass spectrum, researchers can determine the molecular weight of compounds and identify components. With this method, scientists have separated (or isolated) and identified approximately 5000 different chemical compounds. But many of the separated substances are still unidentified, and countless others have not been isolated or studied yet.

3. *Bio-Big Data and Bioinformatics*

Revolutionary advancements in analytical techniques such as next-generation sequencing and UHPLC-HRMS have enabled the analysis of biological data such as DNA sequences, protein sequences, and metabolites. As a result, vast amounts of biological information are available. By leveraging these large datasets, researchers have developed bio-big data databases and used computational systems to analyze this information statistically. Bioinformatics focuses on extracting meaningful insights from such analyses to explain biological phenomena.

Among the disciplines within bioinformatics, genomics and systems biology are particularly notable. Genomics involves analyzing the genome sequences of various organisms to construct databases and study biological evolution and species divergence. Systems biology integrates information from genomics, transcriptomics, proteomics, and metabolomics to connect and explain the interactions underlying biological activities functionally. Through this approach, researchers can model biological systems and assess the accuracy of these models in virtual environments. By applying systems biology analyses to research planning, the success rate of biological experiments can be improved, ultimately saving time and resources. Moreover, systems biology can contribute to the production of valuable compounds for practical applications. For instance, paclitaxel, a widely used chemotherapeutic agent, was initially derived from the bark of the Pacific yew tree. However, extracting paclitaxel from mature trees was unsustainable, as three fully grown trees yielded only about 1 g of the compound. With advancements in technology, researchers could obtain genomic, transcriptomic, proteomic, and metabolomic data from the yew tree, leading to the identification of key genes involved in paclitaxel biosynthesis. By employing heterologous expression of these genes in host organisms such as *Escherichia coli* or tobacco plants, scientists developed methods to produce paclitaxel without harming yew trees.

Bioinformatics is used in ecological research, too. To determine which plant species are present in a given environment, researchers extract DNA from plant-derived materials suspended in the air, such as pollen, petals, leaves, and seeds, and sequence DNA, and compare the sequencing data to existing genomic databases. Compared to traditional ecological studies, which rely on direct observation and specimen collection in the field, this novel approach represents a groundbreaking innovation in ecological research.

4. *Protein Structure Determination*

Artificial intelligence, a leading force in the Fourth Industrial Revolution, has led to major advancements in protein structure determination. Since protein function is largely determined by its structure, understanding protein structures is important. Traditionally, determining protein structure required cloning the gene encoding the protein, producing large quantities of recombinant protein, and purifying and concentrating the protein to form protein crystals. These crystals were then analyzed using synchrotron radiation to obtain X-ray diffraction data, which were further processed to determine the protein structure. Because this process involves multiple complex and technically demanding steps, the number of proteins with resolved structures remains relatively small, even though their amino acid sequences were identified long ago.

However, DeepMind, a subsidiary of Google, recently developed AlphaFold, an artificial intelligence-based program for predicting protein structures. This innovation has made inferring protein structures easier and more accurate. In a benchmark test measuring structural similarity scores, AlphaFold scored 92.4 out of 100, demonstrating an accuracy comparable to traditional X-ray crystallography-based protein structure determination methods. Using AlphaFold, researchers can predict protein structures without undergoing complex experimental procedures. As a result, the mysteries of how various proteins contribute to biological processes are being unraveled at an unprecedented pace. However, since AlphaFold is trained on known protein structures, it may not be capable of accurately predicting the structures of entirely novel proteins that have never been observed before. Additionally, its accuracy remains limited in protein complexes, where multiple proteins interact and form larger structural assemblies.

Epilogues

Youngsook Lee

The initial reason I wanted to write this book was to inform people that plants have social lives, just like we do. However, as I was writing the book, I realized that not only do plants engage in social interactions, but they are central to the social lives of all living beings on Earth. Since plants provide food and habitats to other organisms, all living beings inevitably interact with plants, either directly or indirectly, upon which further social interactions among other living beings are based.

However, plants also need other life forms and depend on social interactions for them to survive. Plants are connected to other plants through their roots, exchanging information and sharing nutrients. Plants not only provide food to microbes, but they also obtain nutrients with the help of these microbes. Moreover, they rely on animals to transfer their reproductive cells and spread their seeds. Thus, various life forms influence the survival and reproduction of plants.

So, what is the most significant factor currently influencing plants? Undoubtedly, it is humans. While microbes and other animals do affect plants, the impact of humans on plants is incomparable. Humans have transformed forests and grasslands into farmlands and cities, domesticated plants for intensive cultivation, and changed the Earth's temperature, forcing plants to adapt to new environmental conditions. Furthermore, many plant species have gone extinct due to human activities, casting a shadow not only on the future of plants but also on that of humanity. Moreover, with the human population surpassing eight billion and continuing to grow, these impacts on plants are likely to persist or even worsen. Does this mean the future of Earth is only bleak?

Throughout human history, there have been similar challenges, and each time, people have successfully overcome these difficulties. For example, in the mid-twentieth century, there were predictions of widespread starvation due to food shortages, but scientists developed high-yielding and disease-resistant wheat and other crops to overcome this. The crisis we face today seems more severe than those of the past, but there is hope to overcome it if we accurately analyze the causes of the crisis and ensure that many people understand them. Currently, the situation

Y. Lee, B. Y. Choi, *Social Lives of Plants*,
https://doi.org/10.1007/978-981-96-9837-0

looks dire, but once people realize that our very lives depend on it, they will come together to restore the Earth's environment, I hope. To restore our ailing Earth's environment, it is necessary to understand the social lives of organisms and create an environment where all life forms can live together in harmony. The foundation for this is to provide plants with the time and space to function effectively at the center of these social interactions.

In the book, I have tried to respect each paper author's interpretation, but I also aimed to point out where there might still be room for different interpretations and where improvements are needed. I believe that as scientists, we must scrutinize the results and interpretations of all papers with a critical eye. Why is skepticism so important? Because people tend to see what they want to see. If everyone insists their own observations are the truth—and we accept them without question—we take in distorted facts. Then, when problems arise, we are less likely to find the right solutions. And if there are no right answers, people will believe in and follow the wrong paths. Scientists have a duty to discern the truth. To uncover the truth, even if an explanation seems plausible, one must continue to doubt and verify whether the evidence is conclusive. I hope that people armed with this scientific attitude will diligently analyze nature and, furthermore, collaborate with each other, to overcome the difficulties humanity faces. It would be a great honor if this book could contribute even slightly in this direction.

Bae Young Choi

On a sunny spring day, dandelions bloomed here and there. Some dandelion flowers stood tall with long stems, while others blossomed with short stems. Normally, I would have simply passed by, but this time, I suddenly recalled the contents of this book and let my imagination run wild. Could it be that the dandelions with long flower stems were employing a strategy to attract more insects to their flowers? Indeed, around the dandelions with long stems, many other tall plants were growing as well. If the dandelions had not extended their stems, their flowers might not have stood out, and I probably would have passed by without seeing them. Interestingly, around the dandelions with short stems, the surrounding grasses were relatively short, too, making dandelion flowers stand out. Through this observation, I hypothesized that perhaps dandelions adjust the length of their flower stems in response to the presence of neighboring plants and compete with them, and the thought made me smile with joy. If we take a closer look at the plants we usually overlook as mere background, you too may begin to imagine the "stories of plants" living wholeheartedly in the ecosystem, just as I did. To borrow a line from a poet Tae-joo Na, the plants we casually overlook are "beautiful when seen up close, and grow more endearing the longer you gaze." Through this, plants can become meaningful neighbors rather than just a background!

In this book, we wrote about how people can observe plants more closely and share the "stories of plants" that scientists have already glimpsed. My hope in writing this book was to encourage readers to take a greater interest in these "stories of plants." Furthermore, I wish for my readers to let their imagination run wild, uncovering the countless untold stories of plants that have not been mentioned in these pages.

Acknowledgments

Youngsook Lee
The idea for this book first took root in the fall of 2016. At the time, I was co-teaching a graduate course at Pohang University of Science and Technology (POSTECH) with Enrico Martinoia, who was visiting from the University of Zurich. The course—Physiological and Chemical Ecology—explored how plants use chemical strategies to survive challenges such as pathogens, herbivores, and heavy metals.

What captivated me most was the section on plant interactions—the ways in which plants compete with and depend on other plants, microorganisms, and animals. These dynamics resemble social behavior, and I began to see striking parallels between plant and human societies. I thought this perspective could make for a compelling book that both entertains and deepens our understanding of nature. It also seemed meaningful to examine issues like global warming through the lens of plant life, offering readers a fresh perspective.

Over the years, I had many valuable conversations about the book's concept and direction with Enrico Martinoia and Kathleen Farquharson, an editor at *The Plant Cell*. In 2019, I launched a graduate course at POSTECH titled The Social Lives of Plants, where I explored these ideas further with a group of bright and curious students. I am grateful to all of them for their contributions.

Bae Young Choi, my co-author and now an assistant professor at Korea National University of Transportation, was one of those students. He played a vital role in shaping this book—contributing text, fact-checking the content against the scientific literature, creating accessible illustrations, adding footnotes for technical terms, and reviewing the manuscript for clarity and balance. I could not have completed this project without his diligent and thoughtful collaboration.

I would also like to thank Sangjoon Park, director of the Institute for Convergence Civilization at POSTECH, for his support and guidance throughout the publication process. Additional thanks go to Joohyun Kang and Yoonyoung Lee (POSTECH) and Yeonsook Choung (Kangwon National University), for their expert advice.

Y. Lee, B. Y. Choi, *Social Lives of Plants*,
https://doi.org/10.1007/978-981-96-9837-0

The many beautiful illustrations and photographs in this book were contributed by Minwoo Oh and Kyungyoon Kim (Seoul National University), Eunwoon Noh (National Institute of Forest Science), Hyunjung Lee (KAIST), Somin Kim, Junsung Park (POSTECH), and other scientists and illustrators who generously provided their valuable work. I am deeply grateful to them.

I am also grateful to Jangryeol Liu, Jongseok Lee, and Sarah Lennon for encouraging the development of this English edition of the book. Special thanks go to Nava Moran and Kathleen Farquharson for their meticulous editing, rigorous feedback, and insightful suggestions.

Bae Young Choi

First of all, I would like to express my deepest gratitude to Professor Youngsook Lee, my co-author and academic advisor, who proposed the idea for this book. She has dedicated her life to exploring the diverse characteristics of plants that cannot move or speak and has always held a profound affection for the greenery that covers our planet. As I wrote this book, I hoped to share her passion for plants with a wider audience. Her interest and devotion to plants were my greatest inspiration throughout the writing process. Working alongside her made me feel a deeper closeness and affection toward these remarkable living beings.

I also extend my heartfelt thanks to Jaewook Kim at Korea National University of Education for his academic advice and to Jae Gyeong Jung at Chungnam National University for meticulously reviewing the manuscript for typos and critical errors. Their contributions greatly enriched and refined the content of this book.

Lastly, I sincerely thank my fiancée, Chayanee Chairattanawat, who stood by my side and offered unwavering support through the many challenges and hardships encountered during the writing process.

References

1. Acevedo FE, Rivera-Vega LJ, Chung SH, Ray S, Felton GW. 2015. Cues from chewing insects—the intersection of DAMPs, HAMPs, MAMPs and effectors. *Current Opinion in Plant Biology* **26**: 80–86.
2. Akiyama K, Matsuzaki K-i, Hayashi H. 2005. Plant sesquiterpenes induce hyphal branching in arbuscular mycorrhizal fungi. *Nature* **435**(7043): 824–827.
3. Alfonso E, Stahl E, Glauser G, Bellani E, Raaymakers TM, Van den Ackerveken G, Zeier J, Reymond P. 2021. Insect eggs trigger systemic acquired resistance against a fungal and an oomycete pathogen. *New Phytologist* **232**(6): 2491–2505.
4. Asseng S, Guarin JR, Raman M, Monje O, Kiss G, Despommier DD, Meggers FM, Gauthier PP. 2020. Wheat yield potential in controlled-environment vertical farms. *Proceedings of the National Academy of Sciences* **117**(32): 19131–19135.
5. Atwood TB, Connolly RM, Ritchie EG, Lovelock CE, Heithaus MR, Hays GC, Fourqurean JW, Macreadie PI. 2015. Predators help protect carbon stocks in blue carbon ecosystems. *Nature Climate Change* **5**(12): 1038–1045.
6. Badri DV, Vivanco JM. 2009. Regulation and function of root exudates. *Plant, Cell & Environment* **32**(6): 666–681.
7. Bahulikar RA, Chaluvadi SR, Torres-Jerez I, Mosali J, Bennetzen JL, Udvardi M. 2021. Nitrogen Fertilization Reduces Nitrogen Fixation Activity of Diverse Diazotrophs in Switchgrass Roots. *Phytobiomes Journal* **5**(1): 80–87.
8. Bakker ES, Gill JL, Johnson CN, Vera FW, Sandom CJ, Asner GP, Svenning J-C. 2016. Combining paleo-data and modern exclosure experiments to assess the impact of megafauna extinctions on woody vegetation. *Proceedings of the National Academy of Sciences* **113**(4): 847–855.
9. Ballaré CL. 2014. Light regulation of plant defense. *Annual Review of Plant Biology* **65**(1): 335–363.
10. Bao T, Wang B, Li J, Dilcher D. 2019. Pollination of Cretaceous flowers. *Proceedings of the National Academy of Sciences* **116**(49): 24707–24711.
11. Berendsen RL, Pieterse CM, Bakker PA. 2012. The rhizosphere microbiome and plant health. *Trends in plant science* **17**(8): 478–486.
12. Borrell JS, Dodsworth S, Forest F, Pérez-Escobar OA, Lee MA, Mattana E, Stevenson PC, Howes MJR, Pritchard HW, Ballesteros D, et al. 2020. The climatic challenge: Which plants will people use in the next century? *Environmental and Experimental Botany* **170**: 103872.
13. Breia R, Conde A, Badim H, Fortes AM, Gerós H, Granell A. 2021. Plant SWEETs: from sugar transport to plant–pathogen interaction and more unexpected physiological roles. *Plant Physiology* **186**(2): 836–852.

Y. Lee, B. Y. Choi, *Social Lives of Plants*,
https://doi.org/10.1007/978-981-96-9837-0

14. Broz A, Vivanco J, Schultz M, Perry L, Paschke M. 2006. Secondary metabolites and allelopathy in plant invasions: A case study of Centaurea maculosa. *Plant Physiology (Essay 13.7). Sinauer Associates.*
15. Burkepile DE, Parker JD. 2017. Recent advances in plant-herbivore interactions. *F1000Research* **6**.
16. Buxton A 2022. Brevel Nets $8.4 Million For Food System Disruption With New Microalgae Alt-Protein. *green queen.*
17. Cascone P, Vuts J, Birkett MA, Dewhirst S, Rasmann S, Pickett JA, Guerrieri E. 2023. L-DOPA functions as a plant pheromone for belowground anti-herbivory communication. *Ecology Letters* **26**(3): 460–469.
18. Castrillo G, Teixeira PJPL, Paredes SH, Law TF, De Lorenzo L, Feltcher ME, Finkel OM, Breakfield NW, Mieczkowski P, Jones CD. 2017. Root microbiota drive direct integration of phosphate stress and immunity. *Nature* **543**(7646): 513–518.
19. Chen H, Wilkerson CG, Kuchar JA, Phinney BS, Howe GA. 2005. Jasmonate-inducible plant enzymes degrade essential amino acids in the herbivore midgut. *Proceedings of the national academy of sciences* **102**(52): 19237–19242.
20. Choi Y-s 2022. [전문가의 눈] 꿀벌 폐사 원인과 대응방안. 농민신문.
21. Choi J, Tanaka K, Cao Y, Qi Y, Qiu J, Liang Y, Lee SY, Stacey G. 2014. Identification of a plant receptor for extracellular ATP. *Science* **343**(6168): 290–294.
22. Colaianni NR, Parys K, Lee H-S, Conway JM, Kim NH, Edelbacher N, Mucyn TS, Madalinski M, Law TF, Jones CD. 2021. A complex immune response to flagellin epitope variation in commensal communities. *Cell Host & Microbe* **29**(4): 635–649. e639.
23. Cook C, Whichard LP, Turner B, Wall ME, Egley GH. 1966. Germination of witchweed (Striga lutea Lour.): isolation and properties of a potent stimulant. *Science* **154**(3753): 1189–1190.
24. Core Writing Team. Intergovernmental Panel on Climate Change (2014) Climate Change 2014: Synthesis Report. Contribution of Working Groups 646 I, II and III to the Fifth Assessment Report of the Intergovernmental Panel on Climate Change; IPCC: Geneva, Switzerland, 2014; p. 15
25. De Moura FF, Moursi M, Donahue Angel M, Angeles-Agdeppa I, Atmarita A, Gironella GM, Muslimatun S, Carriquiry A. 2016. Biofortified β-carotene rice improves vitamin A intake and reduces the prevalence of inadequacy among women and young children in a simulated analysis in Bangladesh, Indonesia, and the Philippines. *The American Journal of Clinical Nutrition* **104**(3): 769–775.
26. Delaux P-M, Schornack S. 2021. Plant evolution driven by interactions with symbiotic and pathogenic microbes. *Science* **371**(6531): eaba6605.
27. Delaux P-M, Radhakrishnan GV, Jayaraman D, Cheema J, Malbreil M, Volkening JD, Sekimoto H, Nishiyama T, Melkonian M, Pokorny L. 2015. Algal ancestor of land plants was preadapted for symbiosis. *Proceedings of the National Academy of Sciences* **112**(43): 13390–13395.
28. Diamond JM. 1998. *Guns, germs and steel: a short history of everybody for the last 13,000 years*: Random House.
29. Doebley JF, Gaut BS, Smith BD. 2006. The Molecular Genetics of Crop Domestication. *Cell* **127**(7): 1309–1321.
30. Dong J, Gruda N, Lam SK, Li X, Duan Z. 2018. Effects of elevated CO2 on nutritional quality of vegetables: a review. *Frontiers in plant science* **9**: 924.
31. Dong Z, Xiao Y, Govindarajulu R, Feil R, Siddoway ML, Nielsen T, Lunn JE, Hawkins J, Whipple C, Chuck G. 2019. The regulatory landscape of a core maize domestication module controlling bud dormancy and growth repression. *Nature Communications* **10**(1): 3810.
32. Douglas AE. 2018. Strategies for enhanced crop resistance to insect pests. *Annual review of plant biology* **69**: 637–660.
33. Dudley SA, File AL. 2007. Kin recognition in an annual plant. *Biology Letters* **3**(4): 435–438.
34. Erb M, Reymond P. 2019. Molecular interactions between plants and insect herbivores. *Annual review of plant biology* **70**: 527–557.

35. Fernie AR, Yan J. 2019. De Novo Domestication: An Alternative Route toward New Crops for the Future. *Molecular Plant* **12**(5): 615–631.
36. Finzi AC, Moore DJ, DeLucia EH, Lichter J, Hofmockel KS, Jackson RB, Kim H-S, Matamala R, McCarthy HR, Oren R. 2006. Progressive nitrogen limitation of ecosystem processes under elevated CO2 in a warm-temperate forest. *Ecology* **87**(1): 15–25.
37. Ford AT, Goheen JR, Otieno TO, Bidner L, Isbell LA, Palmer TM, Ward D, Woodroffe R, Pringle RM. 2014. Large carnivores make savanna tree communities less thorny. *Science* **346**(6207): 346–349.
38. Fragoso V, Rothe E, Baldwin IT, Kim SG. 2014. Root jasmonic acid synthesis and perception regulate folivore-induced shoot metabolites and increase Nicotiana attenuata resistance. *New Phytologist* **202**(4): 1335–1345.
39. Gallé Á, Czékus Z, Tóth L, Galgóczy L, Poór P. 2021. Pest and disease management by red light. *Plant, Cell & Environment* **44**(10): 3197–3210.
40. Gonzales-Vigil E, Bianchetti CM, Phillips Jr GN, Howe GA. 2011. Adaptive evolution of threonine deaminase in plant defense against insect herbivores. *Proceedings of the National Academy of Sciences* **108**(14): 5897–5902.
41. Guo X, Yu Q, Chen D, Wei J, Yang P, Yu J, Wang X, Kang L. 2020. 4-Vinylanisole is an aggregation pheromone in locusts. *Nature* **584**(7822): 584–588.
42. Haack RA, Hérard F, Sun J, Turgeon JJ. 2010. Managing invasive populations of Asian longhorned beetle and citrus longhorned beetle: a worldwide perspective. *Annual review of entomology* **55**: 521–546.
43. Hagihara T, Mano H, Miura T, Hasebe M, Toyota M. 2022. Calcium-mediated rapid movements defend against herbivorous insects in Mimosa pudica. *Nature Communications* **13**(1): 6412.
44. Harada K, Ann JAM, Suzuki M. 2020. Legacy effects of sika deer overpopulation on ground vegetation and soil physical properties. *Forest Ecology and Management* **474**: 118346.
45. Harari YN. 2014. *Sapiens: A brief history of humankind*: Random House.
46. Harris F, Dobbs J, Atkins D, Ippolito JA, Stewart JE. 2021. Soil fertility interactions with Sinorhizobium-legume symbiosis in a simulated Martian regolith; effects on nitrogen content and plant health. *Plos one* **16**(9): e0257053.
47. Hartzler RG. 2010. Reduction in common milkweed (Asclepias syriaca) occurrence in Iowa cropland from 1999 to 2009. *Crop Protection* **29**(12): 1542–1544.
48. Hedden P, Sponsel V. 2015. A century of gibberellin research. *Journal of plant growth regulation* **34**(4): 740–760.
49. Hedrich R, Neher E. 2018. Venus flytrap: how an excitable, carnivorous plant works. *Trends in Plant Science* **23**(3): 220–234.
50. Hettenhausen C, Li J, Zhuang H, Sun H, Xu Y, Qi J, Zhang J, Lei Y, Qin Y, Sun G. 2017. Stem parasitic plant Cuscuta australis (dodder) transfers herbivory-induced signals among plants. *Proceedings of the National Academy of Sciences* **114**(32): E6703–E6709.
51. Hoffland E, Niemann G, Van Pelt J, Pureveen J, Eijkel G, Boon J, Lambers H. 1996. Relative growth rate correlates negatively with pathogen resistance in radish: the role of plant chemistry. *Plant, Cell & Environment* **19**(11): 1281–1290.
52. Hu Y, Ding Y, Cai B, Qin X, Wu J, Yuan M, Wan S, Zhao Y, Xin X-F. 2022. Bacterial effectors manipulate plant abscisic acid signaling for creation of an aqueous apoplast. *Cell Host & Microbe* **30**(4): 518–529. e516.
53. Huang AC, Jiang T, Liu Y-X, Bai Y-C, Reed J, Qu B, Goossens A, Nützmann H-W, Bai Y, Osbourn A. 2019. A specialized metabolic network selectively modulates Arabidopsis root microbiota. *Science* **364**(6440): eaau6389.
54. Hufford MB, Teran JCBMy, Gepts P. 2019. Crop Biodiversity: An Unfinished Magnum Opus of Nature. *Annual Review of Plant Biology* **70**(1): 727–751.
55. Javot H, Penmetsa RV, Terzaghi N, Cook DR, Harrison MJ. 2007. A Medicago truncatula phosphate transporter indispensable for the arbuscular mycorrhizal symbiosis. *Proceedings of the National Academy of Sciences* **104**(5): 1720–1725.

56. Jennings DE, Rohr JR. 2011. A review of the conservation threats to carnivorous plants. *Biological Conservation* **144**(5): 1356–1363.
57. Jeong J, Suh S, Guan C, Tsay Y-F, Moran N, Oh CJ, An CS, Demchenko KN, Pawlowski K, Lee Y. 2004. A nodule-specific dicarboxylate transporter from alder is a member of the peptide transporter family. *Plant Physiology* **134**(3): 969–978.
58. Jiang Y, Wang W, Xie Q, Liu N, Liu L, Wang D, Zhang X, Yang C, Chen X, Tang D. 2017. Plants transfer lipids to sustain colonization by mutualistic mycorrhizal and parasitic fungi. *Science* **356**(6343): 1172–1175.
59. Johnson R, Narvaez J, An G, Ryan C. 1989. Expression of proteinase inhibitors I and II in transgenic tobacco plants: effects on natural defense against Manduca sexta larvae. *Proceedings of the National Academy of Sciences* **86**(24): 9871–9875.
60. Kang S-I 2022. 새로운 GMO, 유전자가위에 맞서며. 한국농정.
61. Kil B-S 1992. Effect of pine allelochemicals on selected species in Korea. *Allelopathy: basic and applied aspects*: Springer, 205–241.
62. Kim K-a 2020. 우리나라 토종종자, 스발바르 국제종자저장고에 영구 보존. *K*스피릿.
63. Kimura F, Sato M, Kato-Noguchi H. 2015. Allelopathy of pine litter: delivery of allelopathic substances into forest floor. *Journal of Plant Biology* **58**: 61–67.
64. Kloth KJ, Busscher-Lange J, Wiegers GL, Kruijer W, Buijs G, Meyer RC, Albrectsen BR, Bouwmeester HJ, Dicke M, Jongsma MA. 2017. SIEVE ELEMENT-LINING CHAPERONE1 restricts aphid feeding on Arabidopsis during heat stress. *The Plant Cell* **29**(10): 2450–2464.
65. Koch GW, Sillett SC, Jennings GM, Davis SD. 2004. The limits to tree height. *Nature* **428**(6985): 851–854.
66. Kodama K, Rich MK, Yoda A, Shimazaki S, Xie X, Akiyama K, Mizuno Y, Komatsu A, Luo Y, Suzuki H. 2022. An ancestral function of strigolactones as symbiotic rhizosphere signals. *Nature communications* **13**(1): 1–15.
67. Konishi S, Izawa T, Lin SY, Ebana K, Fukuta Y, Sasaki T, Yano M. 2006. An SNP Caused Loss of Seed Shattering During Rice Domestication. *Science* **312**(5778): 1392–1396.
68. Koo AJ, Gao X, Daniel Jones A, Howe GA. 2009. A rapid wound signal activates the systemic synthesis of bioactive jasmonates in Arabidopsis. *The Plant Journal* **59**(6): 974–986.
69. Kountche BA, Jamil M, Yonli D, Nikiema MP, Blanco-Ania D, Asami T, Zwanenburg B, Al-Babili S. 2019. Suicidal germination as a control strategy for Striga hermonthica (Benth.) in smallholder farms of sub-Saharan Africa. *Plants, People, Planet* **1**(2): 107–118.
70. Kwak M-J, Kong HG, Choi K, Kwon S-K, Song JY, Lee J, Lee PA, Choi SY, Seo M, Lee HJ. 2018. Rhizosphere microbiome structure alters to enable wilt resistance in tomato. *Nature biotechnology* **36**(11): 1100–1109.
71. Kyndt T, Quispe D, Zhai H, Jarret R, Ghislain M, Liu Q, Gheysen G, Kreuze JF. 2015. The genome of cultivated sweet potato contains *Agrobacterium* T-DNAs with expressed genes: An example of a naturally transgenic food crop. *Proceedings of the National Academy of Sciences* **112**(18): 5844–5849.
72. Lee E-A 2016. 크리스퍼 유전자 교정작물, GMO와 어떻게 다른가. *BIOSPECTATOR*.
73. Legg S. 2021. IPCC, 2021: Climate Change 2021-the Physical Science basis. *Interaction* **49**(4): 44–45.
74. Li K-T, Moulin M, Mangel N, Albersen M, Verhoeven-Duif NM, Ma Q, Zhang P, Fitzpatrick TB, Gruissem W, Vanderschuren H. 2015. Increased bioavailable vitamin B6 in field-grown transgenic cassava for dietary sufficiency. *Nature Biotechnology* **33**(10): 1029–1032.
75. Lin Z, Eaves DJ, Sanchez-Moran E, Franklin FCH, Franklin-Tong VE. 2015. The Papaver rhoeas S determinants confer self-incompatibility to Arabidopsis thaliana in planta. *Science* **350**(6261): 684–687.
76. Little D, Gouhier-Darimont C, Bruessow F, Reymond P. 2007. Oviposition by pierid butterflies triggers defense responses in Arabidopsis. *Plant Physiology* **143**(2): 784–800.
77. Liu Z, Hou S, Rodrigues O, Wang P, Luo D, Munemasa S, Lei J, Liu J, Ortiz-Morea FA, Wang X. 2022. Phytocytokine signalling reopens stomata in plant immunity and water loss. *Nature* **605**(7909): 332–339.

78. López Pereira M, Sadras VO, Batista W, Casal JJ, Hall AJ. 2017. Light-mediated self-organization of sunflower stands increases oil yield in the field. *Proceedings of the National Academy of Sciences* **114**(30): 7975–7980.
79. Lortzing V, Oberländer J, Lortzing T, Tohge T, Steppuhn A, Kunze R, Hilker M. 2019. Insect egg deposition renders plant defence against hatching larvae more effective in a salicylic acid-dependent manner. *Plant, Cell & Environment* **42**(3): 1019–1032.
80. Mallik AU. 2008. Allelopathy in forested ecosystems. *Allelopathy in sustainable agriculture and forestry*: 363–386.
81. Manzano A, Herranz R, den Toom LA, Te Slaa S, Borst G, Visser M, Medina FJ, van Loon JJ. 2018. Novel, Moon and Mars, partial gravity simulation paradigms and their effects on the balance between cell growth and cell proliferation during early plant development. *npj Microgravity* **4**(1): 1–11.
82. Melillo JM, Butler S, Johnson J, Mohan J, Steudler P, Lux H, Burrows E, Bowles F, Smith R, Scott L, et al. 2011. Soil warming, carbon-nitrogen interactions, and forest carbon budgets. *Proceedings of the National Academy of Sciences* **108**(23): 9508–9512.
83. Mendes LW, Raaijmakers JM, de Hollander M, Mendes R, Tsai SM. 2018. Influence of resistance breeding in common bean on rhizosphere microbiome composition and function. *The ISME journal* **12**(1): 212–224.
84. Meyer RS, DuVal AE, Jensen HR. 2012. Patterns and processes in crop domestication: an historical review and quantitative analysis of 203 global food crops. *New Phytologist* **196**(1): 29–48.
85. Moog MW, Trinh MDL, Nørrevang AF, Bendtsen AK, Wang C, Østerberg JT, Shabala S, Hedrich R, Wendt T, Palmgren M. 2022. The epidermal bladder cell-free mutant of the salt-tolerant quinoa challenges our understanding of halophyte crop salinity tolerance. *New Phytologist* **236**(4): 1409–1421.
86. National Academies of Sciences Engineering Medicine. 2016. *Genetically engineered crops: experiences and prospects*: National Academies Press.
87. National Institute of Biological Resources 2012. Red data book of endangered vascular plants in Korea: National Institute of Biological Resources Incheon.
88. National Research Council. 2007. *Status of Pollinators in North America*. Washington, DC: The National Academies Press.
89. Nelsen MP, Moreau CS, Kevin Boyce C, Ree RH. 2023. Macroecological diversification of ants is linked to angiosperm evolution. *Evolution Letters* **7**(2): 79–87.
90. Nomura K, Andreazza F, Cheng J, Dong K, Zhou P, He SY. 2023. Bacterial pathogens deliver water-and solute-permeable channels to plant cells. *Nature* **621**(7979): 586–591.
91. Normile D 2010. Restoration or devastation?: American Association for the Advancement of Science.
92. Oliva N, Florida Cueto-Reaño M, Trijatmiko KR, Samia M, Welsch R, Schaub P, Beyer P, Mackenzie D, Boncodin R, Reinke R, et al. 2020. Molecular characterization and safety assessment of biofortified provitamin A rice. *Scientific Reports* **10**(1): 1376.
93. Ollerton J, Winfree R, Tarrant S. 2011. How many flowering plants are pollinated by animals? *Oikos* **120**(3): 321–326.
94. Orrock J, Connolly B, Kitchen A. 2017. Induced defences in plants reduce herbivory by increasing cannibalism. *Nature ecology & evolution* **1**(8): 1205–1207.
95. Ostiguy N. 2011. Pests and pollinators. *Nature Education Knowledge* **3**(10): 3.
96. Paine JA, Shipton CA, Chaggar S, Howells RM, Kennedy MJ, Vernon G, Wright SY, Hinchliffe E, Adams JL, Silverstone AL, et al. 2005. Improving the nutritional value of Golden Rice through increased pro-vitamin A content. *Nature Biotechnology* **23**(4): 482–487.
97. Park S 2012. 엽채류 해충 천적으로 퇴치. 농민신문.
98. Paul A-L, Elardo SM, Ferl R. 2022. Plants grown in Apollo lunar regolith present stress-associated transcriptomes that inform prospects for lunar exploration. *Communications biology* **5**(1): 1–9.
99. Peakall R. 2023. Pollination by sexual deception. *Current Biology* **33**(11): R489–R496.

100. Pendergast IV TH, Hanlon SM, Long ZM, Royo AA, Carson WP. 2016. The legacy of deer overabundance: long-term delays in herbaceous understory recovery. *Canadian Journal of Forest Research* **46**(3): 362–369.
101. Pickles BJ, Wilhelm R, Asay AK, Hahn AS, Simard SW, Mohn WW. 2017. Transfer of 13C between paired Douglas-fir seedlings reveals plant kinship effects and uptake of exudates by ectomycorrhizas. *New Phytologist* **214**(1): 400–411.
102. Poore J, Nemecek T. 2018. Reducing food's environmental impacts through producers and consumers. *Science* **360**(6392): 987–992.
103. Pourcel L, Moulin M, Fitzpatrick TB. 2013. Examining strategies to facilitate vitamin B1 biofortification of plants by genetic engineering. *Frontiers in plant science* **4**: 160.
104. Puginier C, Keller J, Delaux P-M. 2022. Plant–microbe interactions that have impacted plant terrestrializations. *Plant Physiology* **190**(1): 72–84.
105. Rich MK, Vigneron N, Libourel C, Keller J, Xue L, Hajheidari M, Radhakrishnan GV, Le Ru A, Diop SI, Potente G. 2021. Lipid exchanges drove the evolution of mutualism during plant terrestrialization. *Science* **372**(6544): 864–868.
106. Ritchie H, Roser M. 2020. Environmental impacts of food production. *Our world in data.*
107. Ritchie H, Roser M, Rosado P. 2020. CO_2 and greenhouse gas emissions. *Our world in data.*
108. Román-Palacios C, Wiens JJ. 2020. Recent responses to climate change reveal the drivers of species extinction and survival. *Proceedings of the National Academy of Sciences* **117**(8): 4211–4217.
109. Roussin-Léveillée C, Lajeunesse G, St-Amand M, Veerapen VP, Silva-Martins G, Nomura K, Brassard S, Bolaji A, He SY, Moffett P. 2022. Evolutionarily conserved bacterial effectors hijack abscisic acid signaling to induce an aqueous environment in the apoplast. *Cell Host & Microbe* **30**(4): 489–501. e484.
110. Ruan J, Zhou Y, Zhou M, Yan J, Khurshid M, Weng W, Cheng J, Zhang K. 2019. Jasmonic acid signaling pathway in plants. *International journal of molecular sciences* **20**(10): 2479.
111. Sasse J, Schlegel M, Borghi L, Ullrich F, Lee M, Liu GW, Giner JL, Kayser O, Bigler L, Martinoia E. 2016. Petunia hybrida PDR2 is involved in herbivore defense by controlling steroidal contents in trichomes. *Plant, cell & environment* **39**(12): 2725–2739.
112. Savary S, Willocquet L, Pethybridge SJ, Esker P, McRoberts N, Nelson A. 2019. The global burden of pathogens and pests on major food crops. *Nature ecology & evolution* **3**(3): 430–439.
113. Schiestl FP, Peakall R, Mant JG, Ibarra F, Schulz C, Franke S, Francke W. 2003. The chemistry of sexual deception in an orchid-wasp pollination system. *Science* **302**(5644): 437–438.
114. Schultheiss P, Nooten SS, Wang R, Wong MK, Brassard F, Guénard B. 2022. The abundance, biomass, and distribution of ants on Earth. *Proceedings of the National Academy of Sciences* **119**(40): e2201550119.
115. Semchenko M, Saar S, Lepik A. 2014. Plant root exudates mediate neighbour recognition and trigger complex behavioural changes. *New Phytologist* **204**(3): 631–637.
116. Semchenko M, Xue P, Leigh T. 2021. Functional diversity and identity of plant genotypes regulate rhizodeposition and soil microbial activity. *New Phytologist* **232**(2): 776–787.
117. Shen G, Liu N, Zhang J, Xu Y, Baldwin IT, Wu J. 2020. Cuscuta australis (dodder) parasite eavesdrops on the host plants' FT signals to flower. *Proceedings of the National Academy of Sciences* **117**(37): 23125–23130.
118. Shiojiri K, Kishimoto K, Ozawa R, Kugimiya S, Urashimo S, Arimura G, Horiuchi J, Nishioka T, Matsui K, Takabayashi J. 2006. Changing green leaf volatile biosynthesis in plants: an approach for improving plant resistance against both herbivores and pathogens. *Proceedings of the National Academy of Sciences* **103**(45): 16672–16676.
119. Shulaev V, Silverman P, Raskin I. 1997. Airborne signalling by methyl salicylate in plant pathogen resistance. *Nature* **385**(6618): 718–721.
120. Simard SW, Perry DA, Jones MD, Myrold DD, Durall DM, Molina R. 1997. Net transfer of carbon between ectomycorrhizal tree species in the field. *Nature* **388**(6642): 579–582.
121. Simard SW, Beiler KJ, Bingham MA, Deslippe JR, Philip LJ, Teste FP. 2012. Mycorrhizal networks: mechanisms, ecology and modelling. *Fungal Biology Reviews* **26**(1): 39–60.

122. Song YY, Simard SW, Carroll A, Mohn WW, Zeng RS. 2015. Defoliation of interior Douglas-fir elicits carbon transfer and stress signalling to ponderosa pine neighbors through ectomycorrhizal networks. *Scientific Reports* **5**(1): 8495.
123. Soroye P, Newbold T, Kerr J. 2020. Climate change contributes to widespread declines among bumble bees across continents. *Science* **367**(6478): 685–688.
124. Stringlis IA, Yu K, Feussner K, de Jonge R, Van Bentum S, Van Verk MC, Berendsen RL, Bakker PA, Feussner I, Pieterse CM. 2018. MYB72-dependent coumarin exudation shapes root microbiome assembly to promote plant health. *Proceedings of the National Academy of Sciences* **115**(22): E5213–E5222.
125. Studer A, Zhao Q, Ross-Ibarra J, Doebley J. 2011. Identification of a functional transposon insertion in the maize domestication gene tb1. *Nature Genetics* **43**(11): 1160–1163.
126. Sun J, Lu M, Gillette NE, Wingfield MJ. 2013. Red turpentine beetle: innocuous native becomes invasive tree killer in China. *Annual Review of Entomology* **58**: 293–311.
127. Teixeira PJ, Colaianni NR, Law TF, Conway JM, Gilbert S, Li H, Salas-González I, Panda D, Del Risco NM, Finkel OM. 2021. Specific modulation of the root immune system by a community of commensal bacteria. *Proceedings of the National Academy of Sciences* **118**(16): e2100678118.
128. Thorén LM, Karlsson PS. 1998. Effects of supplementary feeding on growth and reproduction of three carnivorous plant species in a subarctic environment. *Journal of Ecology* **86**(3): 501–510.
129. Tooker JF, Frank SD. 2012. Genotypically diverse cultivar mixtures for insect pest management and increased crop yields. *Journal of Applied Ecology* **49**(5): 974–985.
130. Torices R, Gómez JM, Pannell JR. 2018. Kin discrimination allows plants to modify investment towards pollinator attraction. *Nature Communications* **9**(1): 2018.
131. Trivedi P, Leach JE, Tringe SG, Sa T, Singh BK. 2020. Plant–microbiome interactions: from community assembly to plant health. *Nature reviews microbiology* **18**(11): 607–621.
132. Tuysuz G, Damon A 2015. Arctic 'Doomsday Vault' opens to retrieve vital seeds for Syria. *CNN*.
133. van der Heijden MGA, Martin FM, Selosse MA, Sanders IR. 2015. Mycorrhizal ecology and evolution: the past, the present, and the future. *New phytologist* **205**(4): 1406–1423.
134. Wamelink GW, Frissel JY, Krijnen WH, Verwoert MR, Goedhart PW. 2014. Can plants grow on Mars and the moon: a growth experiment on Mars and moon soil simulants. *PLoS One* **9**(8): e103138.
135. Whitham TG, Bailey JK, Schweitzer JA, Shuster SM, Bangert RK, LeRoy CJ, Lonsdorf EV, Allan GJ, DiFazio SP, Potts BM. 2006. A framework for community and ecosystem genetics: from genes to ecosystems. *Nature Reviews Genetics* **7**(7): 510–523.
136. Wieder WR, Cleveland CC, Smith WK, Todd-Brown K. 2015. Future productivity and carbon storage limited by terrestrial nutrient availability. *Nature Geoscience* **8**(6): 441–444.
137. Wirth JP, Petry N, Tanumihardjo SA, Rogers LM, McLean E, Greig A, Garrett GS, Klemm RDW, Rohner F. 2017. Vitamin A Supplementation Programs and Country-Level Evidence of Vitamin A Deficiency. *Nutrients* **9**(3): 190.
138. Xin X-F, Nomura K, Aung K, Velásquez AC, Yao J, Boutrot F, Chang JH, Zipfel C, He SY. 2016. Bacteria establish an aqueous living space in plants crucial for virulence. *Nature* **539**(7630): 524–529.
139. Xu K, Mackill DJ. 1996 A Major Locus for Submergence Tolerance Mapped on Rice Chromosome 9. *Molecular Breeding*, **2**, 219–224.
140. Xu K, Xu X, Fukao T, Canlas P, Maghirang-Rodriguez R, Heuer S, Ismail AM, Bailey-Serres J, Ronald PC, Mackill DJ. 2006 Sub1A is an ethylene-response-factor-like gene that confers submergence tolerance to rice. *Nature* **10**;442(7103):705–8.
141. Yu H, Li J. 2022. Breeding future crops to feed the world through de novo domestication. *Nature Communications* **13**(1): 1–4.
142. Yu H, Lin T, Meng X, Du H, Zhang J, Liu G, Chen M, Jing Y, Kou L, Li X. 2021. A route to de novo domestication of wild allotetraploid rice. *Cell* **184**(5): 1156–1170. e1114.
143. Zhang M, Kong X. 2021. How plants discern friends from foes. *Trends in Plant Science*.

144. Zhang Q, Blaylock LA, Harrison MJ. 2010. Two Medicago truncatula half-ABC transporters are essential for arbuscule development in arbuscular mycorrhizal symbiosis. *The Plant Cell* **22**(5): 1483–1497.
145. Zhang J, Liu Y-X, Zhang N, Hu B, Jin T, Xu H, Qin Y, Yan P, Zhang X, Guo X. 2019. NRT1. 1B is associated with root microbiota composition and nitrogen use in field-grown rice. *Nature biotechnology* **37**(6): 676–684.
146. Zhang C, He J, Dai H, Wang G, Zhang X, Wang C, Shi J, Chen X, Wang D, Wang E. 2021. Discriminating symbiosis and immunity signals by receptor competition in rice. *Proceedings of the National Academy of Sciences* **118**(16): e2023738118.
147. Zhou F, Emonet A, Tendon VD, Marhavy P, Wu D, Lahaye T, Geldner N. 2020. Co-incidence of damage and microbial patterns controls localized immune responses in roots. *Cell* **180**(3): 440–453. e418.
148. Ziska LH, Epstein PR, Schlesinger WH. 2009. Rising CO2, Climate Change, and Public Health: Exploring the Links to Plant Biology. *Environmental Health Perspectives* **117**(2): 155–158.
149. Zsögön A, Čermák T, Naves ER, Notini MM, Edel KH, Weinl S, Freschi L, Voytas DF, Kudla J, Peres LEP. 2018. De novo domestication of wild tomato using genome editing. *Nature biotechnology* **36**(12): 1211–1216.

MIX
Papier aus verantwortungsvollen Quellen
Paper from responsible sources
FSC® C105338

If you have any concerns about our products,
you can contact us on
ProductSafety@springernature.com

In case Publisher is established outside the EU,
the EU authorized representative is:
Springer Nature Customer Service Center GmbH
Europaplatz 3, 69115 Heidelberg, Germany

Printed by Libri Plureos GmbH
in Hamburg, Germany